Sustainable Food Production 2014

Topical Conference at the 2014 AIChE Annual Meeting

Atlanta, Georgia, USA
16-21 November 2014

ISBN: 978-1-5108-1249-9

Printed from e-media with permission by:

Curran Associates, Inc.
57 Morehouse Lane
Red Hook, NY 12571

proceedings.com

Some format issues inherent in the e-media version may also appear in this print version.

Printed by Curran Associates, Inc. (2015)

AIChE
120 Wall Street, FL 23
New York, NY 10005-4020

Phone: (800) 242-4363
Fax: (203) 775-5177

www.aiche.org

Additional copies of this publication are available from:

Curran Associates, Inc.
57 Morehouse Lane
Red Hook, NY 12571 USA
Phone: 845-758-0400
Fax: 845-758-2634
Email: curran@proceedings.com
Web: www.proceedings.com

TABLE OF CONTENTS

364033 Self-Emulsification of Alkaline-Dissolved Clove Bud Oil By Whey Protein, Gum Arabic, Lecithin, and Their Combinations

Abstract:

Essential oils (EOs) are well-known for their antioxidant and antimicrobial activities and have been widely used as functional ingredients in food, pharmaceutical and cosmetic products. However, poor water solubility and high volatility during processing are two major obstacles for utilizing EOs as preservatives in food matrices or as sanitizing agents. In the present study, a novel self-emulsifying process was studied to prepare clove bud oil (CBO) emulsion without any specialized equipment or organic solvents. In this method, the major CBO component (eugenol) was deprotonated at alkaline conditions, which was accelerated by heating. Heating 10% v/v CBO in 3 M NaOH at 110 °C for 10 min was observed to be optimal conditions to obtain a transparent CBO solution. The alkaline-dissolved CBO was mixed at an overall CBO concentration of 1% v/v in aqueous neutral solutions containing 1% w/v naturally-occurring emulsifiers - whey protein concentrate (WPC), gum arabic, lecithin, and their combinations. Emulsions were obtained after neutralization to pH 7, during which eugenol became re-protonated and encapsulated in situ by emulsifiers. CBO extracted from emulsions had identical UV-visible absorption spectrum as native CBO, and the peak in reversed-phase HPLC had a same elution time and peak area as the native eugenol. These results verified the maintained structure of eugenol after self-emulsification. The freshly prepared emulsions had hydrodynamic diameters in the range of 160–540 nm, but their dimension and stability were dependent on the emulsifiers. The sample self-emulsified with WPC only had the smallest particles, followed by the combination of all three emulsifiers and the WPC-lecithin mixture, while that with gum arabic alone had the largest particles. The emulsifier blend generally resulted in better stability of the emulsion particle size than individual compounds. Emulsions were evaluated for their antimicrobial activities against *Escherichia coli* O157:H7 ATCC 43895, *Listeria monocytogenes* Scott A and *Salmonella* Enteritidis. Emulsions generally exhibited better activity than same amounts of ethanol-dissolved CBO, showing lower minimum inhibitory and bactericidal concentrations and stronger inhibition of bacterial growth in tryptic soy broth. Our findings indicated that this novel self-emulsifying process may find promising applications in preparing EO emulsions using low cost food ingredients for use as antimicrobial food preservatives.

OIL EXTRACTION FROM SPENT COFFEE GROUNDS USING ADVANCED TECHNIQUES

Javier A. Dávila, Moshe Rosenberg, Gonzalo Taborda, Carlos A. Cardona
Universidad Nacional de Colombia, Manizales. University of California, Davis, Universidad de Caldas

2014 AIChe Annual Meeting

1. INTRODUCTION

SPENT COFFEE GROUNDS (SCG)

Coffee is one of the most important agroindustrial products in Colombia generating appreciable amounts of residues

654,000 T/Year

- Accounts the 10% of the total weight of the grain
- Contains bioactive compounds (Chlorogenic acids)
- Oil content (4% - 25%)
- 44% of fiber (Cellulose, hemicellulose)

2

1. INTRODUCTION

SPENT COFFEE GROUNDS (SCG)

<u>Products</u>

- Cellulose
- Hemicellulose

- Alcohols
- Biofuels
- Bioproducts
- Furfural
- PHB
- Xylitol
- Acids

OIL

- Food
- Pharmaceutical
- Biofuels

(Rodriguez, Cenicafe, 2011)

1. INTRODUCTION

CHEMICAL COMPOSITION OF SCG

Component	(%)
Nitrogen	1.64
Protein	10.30
Ash	0.59
Fiber (Hollocelulose)	39.72
Lipids	25.83
Extractives	21.92

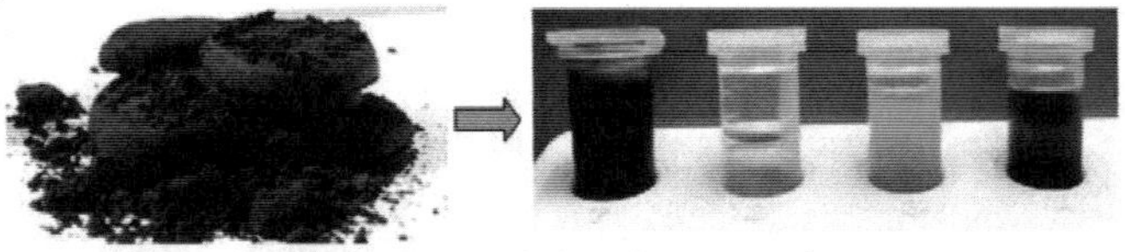

Value Added Product

1. INTRODUCTION

EXTRACTION TECHNIQUES

Most common techniques used:

- Mechanical extraction (High pressure)
- Solvent extraction (Hexane)
- Supercritical extraction (CO2 at supercritical conditions)

6

1. INTRODUCTION

HOW TO EXTRACT OIL FROM SCG

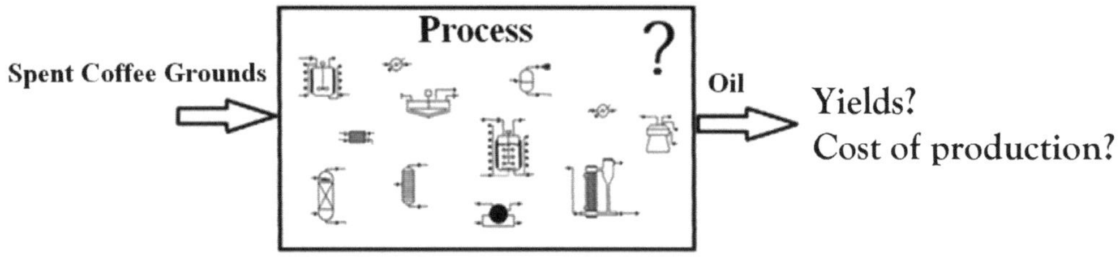

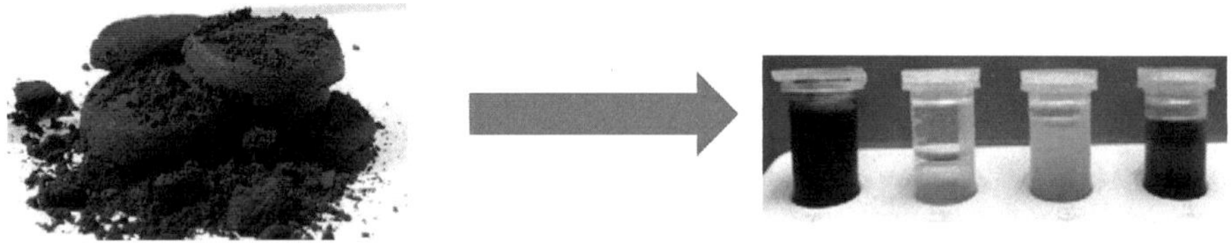

7

1. INTRODUCTION

INTEREST OF THIS WORK

Oil from Spent Coffee Grounds

Motivation:

Extract oil from SCG using three different techniques and obtain a preliminary techno-economic analysis of the extractions

2. METODOLOGY

METHODOLOGY

Experimental

- Mechanical extraction
- Solvent extraction
- Supercritical extraction

Process Simulation

- Construction of PFD
- Mass and energy balances

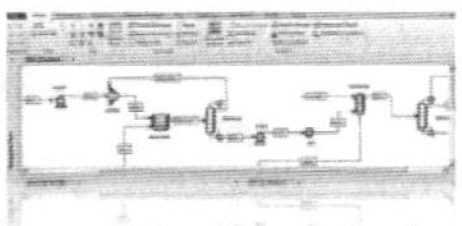

Techno-economic Analysis

- Economic parameters (Colombian context)
- Economic evaluation

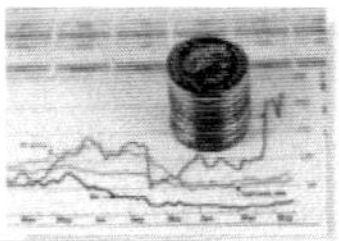

2. METODOLOGY

MECHANICAL EXTRACTION

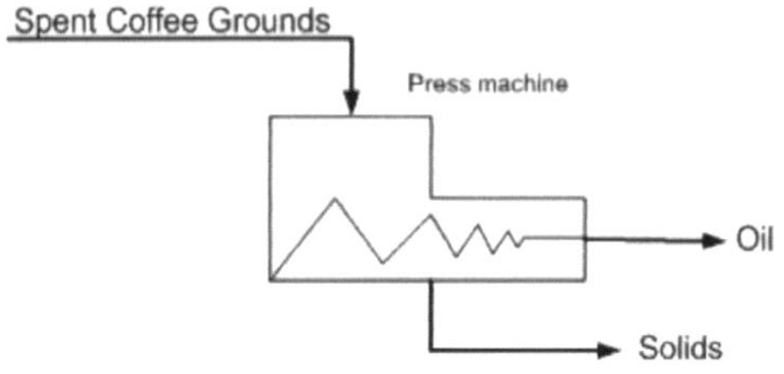

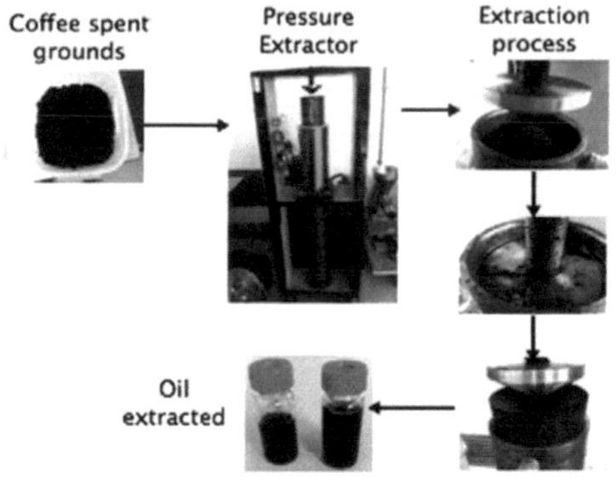

MECHANICAL EXTRACTION

- Pressure: 60 bar
- Temperature: 50, 70, 90 °C

2. METODOLOGY

SOLVENT EXTRACTION

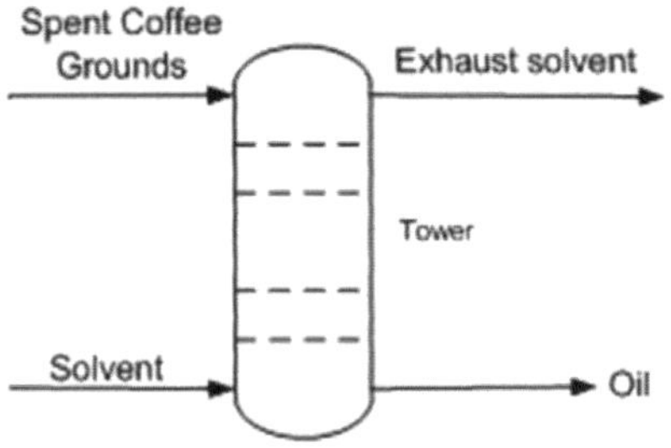

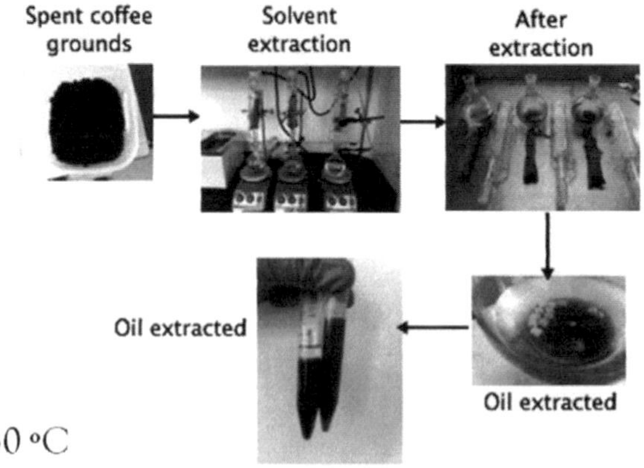

SOLVENT EXTRACTION

- Solvent: Hexane
- Temperature: 90, 120, 160 °C
- Time: 24 h

2. METODOLOGY

SUPERCRITICAL EXTRACTION

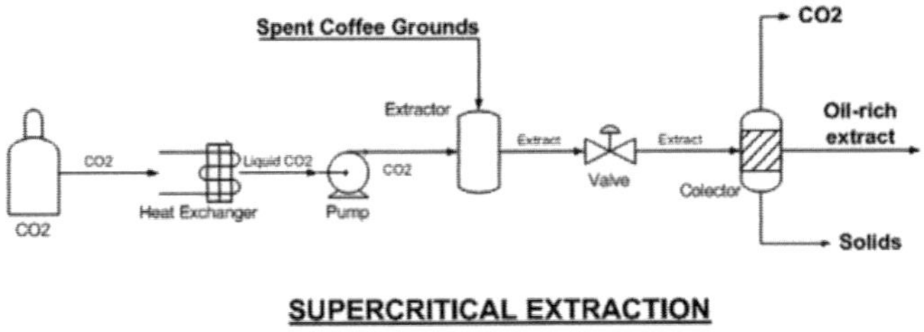

Supercritical Extraction

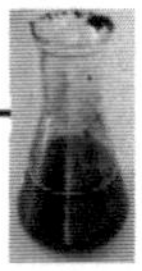

- Solvent: $CO2$
- Temperature: 45 °C, pressure 300 bar
- Time: 5 h

2. METODOLOGY

PROCESS SIMULATION

Characterization of SCG → Simulation of the extraction processes → Mass and Energy Balances (ASPEN PLUS v8.0 software) → Economic Evaluation (Aspen Process Economic Analyzer) → Flows, yield, energy and cost results → Analysis of the process

3. RESULTS

EXPERIMENTAL RESULTS

Yields

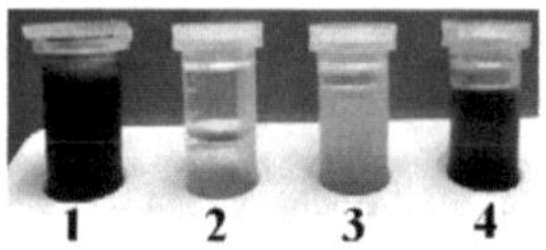

Oil 1: 60 bar, 50, 70, 90 °C
Oil 2: Hexane, 90 °C
Oil 3: Hexane, 120 or 160 °C
Oil 4: Hexane, 120 or 160 °C evaporating inmediately.
Supercritical extraction

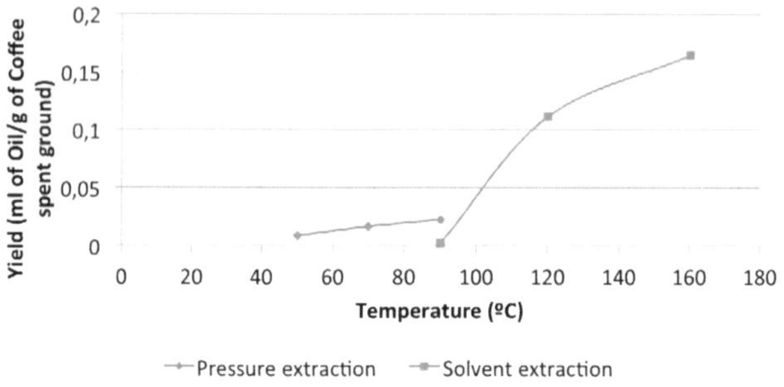

Melanoidins
Liposoluble dyes

Technology	Oil (g/g)	Yield (%)
Solvent extraction	0.165	63.88
Mechanical extraction	0.02	8.96
Supercritical extraction	0.2066	80

3. RESULTS

ECONOMIC ANALYSIS

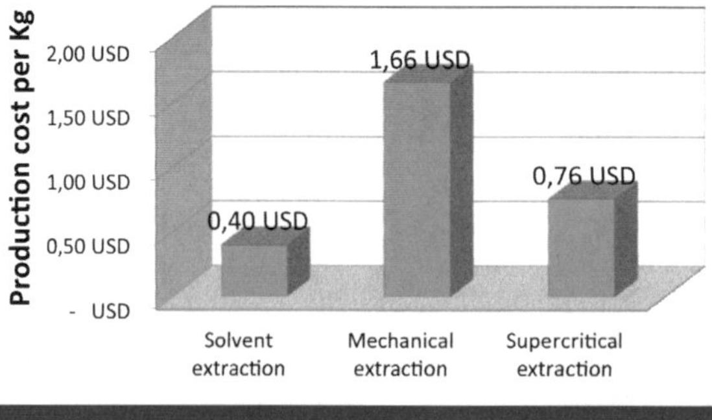

Market price of Palm Oil 0.64 USD/kg

(Index Mundi, 2014)

3. RESULTS

ECONOMIC ANALYSIS

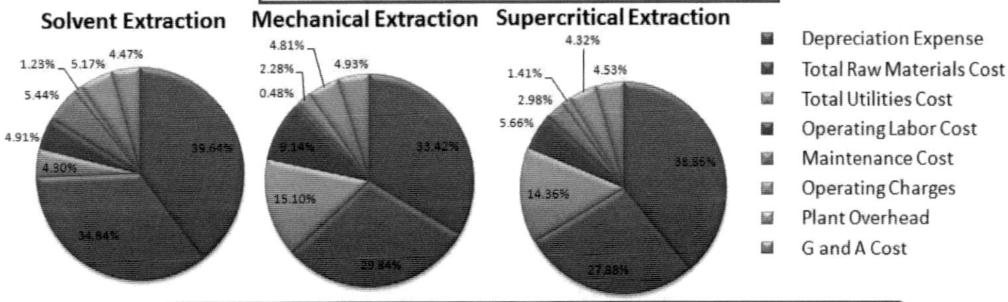

PRODUCTION COST PER YEAR

Solvent Extraction (1.048 million USD/Year)

Mechanical Extraction (0.617 million USD/Year)

Supercritical Extraction (1.211 million USD/Year)

3. RESULTS

OIL EXTRACTED

Oil extracted from SCG

Free Fatty Acids (FFA) were measured according to $AOCS\ Ca\ 5a-40$ official method (Rukunudin et al., 1998)

Technology	Acid Index (% FFA)
Solvent extraction	3.26
Mechanical extraction	3.08
Supercritical extraction	2.95
Palm Oil	2.4 (Rincón et al., 2009)
Coffee Oil from SCG	2.99 (Ramírez, 2008)

Acid Index values ranges from 0.1 % to 4.7 % for biodiesel production.

4. CONCLUSIONS

CONCLUSIONS

1. Oil extracted from SCG is a potential raw material to be for other purposes.
2. Solvent extraction is the most interest technique to obtain oil from SCG.
3. Acid Index indicates that oil from SCG can to be used for biodiesel production.

18

REFERENCES

REFERENCES

1. Rodriguez Valencia Nelson. 2011. Recent experiences on the use of byproducts from coffee. CENICAFE 2011.
2. Price Indexes. Palm oil price market. 2014. Available in www:indexmundi:com
3. Rincon M. Sandra Milena, Martinez C. Daniel Mauricio. Analysis of the properties of palm oil in the industries development. 2009. Palmas Vol 30 No. 2.
4. Ramirez Ovalle Lourdes Mara. Assessment of the yield extraction and characterization of oil from coffee. 2008. Bachelor thesis, Guatemala Universidad de San Carlos de Guatemala.

19

Microencapsulation of Blackberry antioxidants. Modeling and simulation

Miguel A. Rojas[1], Javier A. Davila[1] Moshe Rosenberg[2], Carlos A. Cardona[1*]

[1] Biotechnology and Agribusiness Institute, Department of Chemical Engineering
Universidad Nacional de Colombia Sede Manizales, Manizales -Colombia
[2] University of California Davis, Food Science & Technology Department

marojasre@unal.edu.co; jadavilar@unal.edu.co;
mrosenberg@ucdavis.edu; *ccardonaal@unal.edu.co;

14AIChE
Annual Meeting, Atlanta, GA

1. INTRODUCTION

The Interest in Designing functional foods

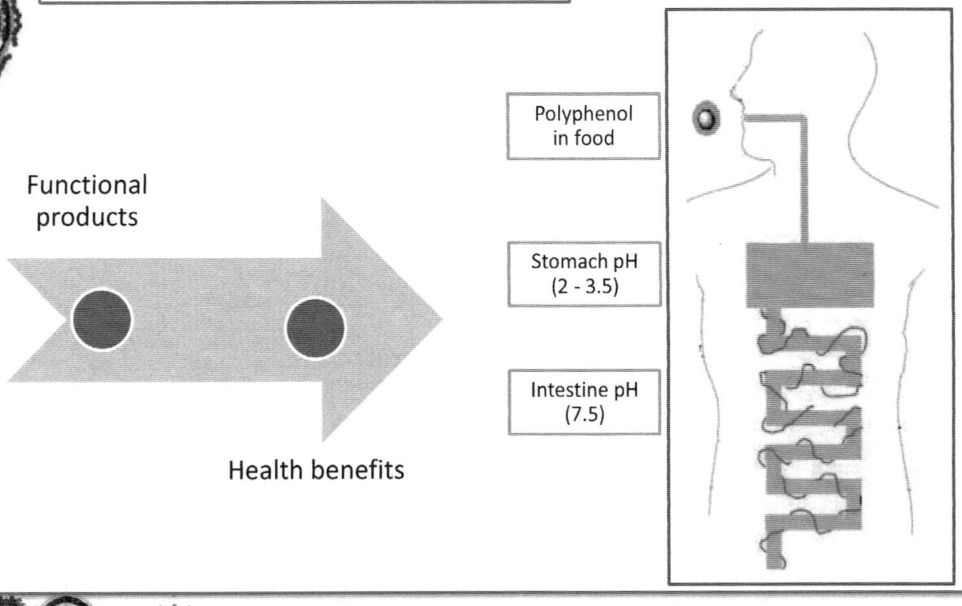

1. INTRODUCTION

Foods rich in polyphenols

- Red Wine
- Green tea
- Coffee
- Pomegranate
- **Blackberry**
- Nuts
- Prune
- Cholocate

Development of natural products with high contents of benefical compounds

Significant market for functional foods

14AIChE Annual Meeting, Atlanta, GA

3

Research Group in Chemical, Catalytic and Biotechnological Processes
Universidad Nacional de Colombia - Sede Manizales

1. INTRODUCTION

Foods rich in polyphenols

- Red Wine
- Green tea
- Coffee
- Pomegranate
- **Blackberry**
- Nuts
- Prune
- Cholocate

Blackberry (*Rubus glaucus benth*) is an exotic tropical fruit that have high content of natural antioxidants, phenolic compounds and anthocyanins.

In 2013, Colombia produced 105,285 tonnes of blackberry from 11,986 ha.

14AIChE Annual Meeting, Atlanta, GA

4

Research Group in Chemical, Catalytic and Biotechnological Processes
Universidad Nacional de Colombia - Sede Manizales

1. INTRODUCTION

After the extraction of active ingredients they need to be immediately protected against biological activity

Food manufacturers

- Polyphenols
- Antimicrobial
- Flavors
- Preservatives

Problems

Chemically and physically unstable

They are incompatible with the product matrix

Lack the appropriate functional attributes

14AIChE Annual Meeting, Atlanta, GA

5

Research Group in Chemical, Catalytic and Biotechnological Processes
Universidad Nacional de Colombia - Sede Manizales

1. INTRODUCTION

After the extraction of active ingredients they need to be immediately protected against biological activity

Food manufacturers

- Polyphenols
- Antimicrobial
- Flavors
- Preservatives

Challenges

Overcome by incorporating the active ingredients into some kind of delivery system before it is introduced into the final product

14AIChE Annual Meeting, Atlanta, GA

6

Research Group In Chemical, Catalytic and Biotechnological Processes
Universidad Nacional de Colombia - Sede Manizales

1. INTRODUCTION

Microencapsulation is a potential solution to the loss of biological activity and maintain the stability of polyphenols during food processing and store

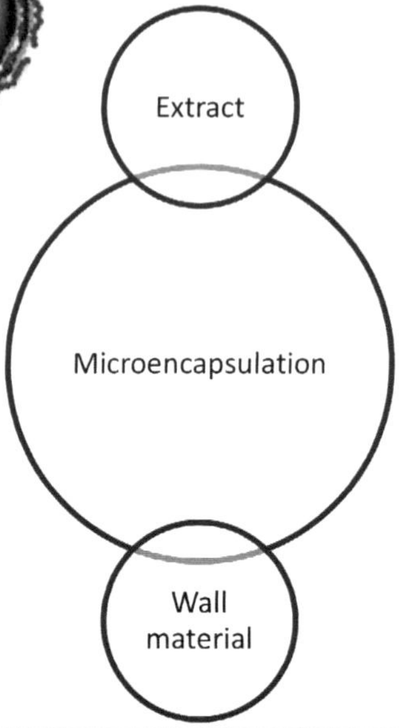

Shell Materials used for microencapsulation purposes

- Polysaccharides: Maltodextrin, Dextrin, Cyclodextrin, Cellulose, gum Arabic, Alginate, Pectin
- Protein: Soy, Wheat, Corn, Gelatin, Casein, Whey protein Isolate, Caseinate
- Polymers: Chitosan, Ethyl cellulose, Polyethylene glycol
- Lipids: Hard fat, Glycerides, Phospholipids
- Waxes: Beesswax, carnuba wax

1. INTRODUCTION

Microencapsulation is a potential solution to the loss of biological activity and maintain the stability of polyphenols during food processing and store

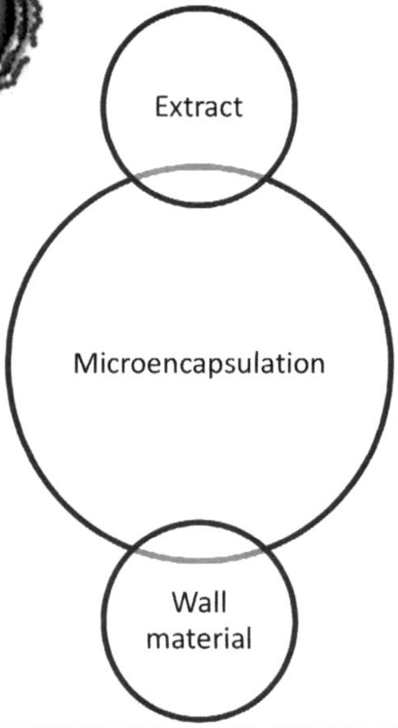

Microencapsulation methods used within the food industry.

Physical	Chemical
Spray drying	Phase separation
Spray cooling	Solvent evaporation
Extrusion	Interfacial polarization
Coextrusion	Liposomes

2. OBJECTIVE

- The objective of this study was to evaluate the influence of microencapsulation by spray drying of blackberry extract on the antioxidant activity and phenolic content and also to obtain the economic analysis of the process

3. METHODOLOGY

Extraction Process → Sample Preparation → Spray Drying → Chemical Analysis → Modeling and Simulation → Economic Evaluation

Supercritical fluid extraction

This Blackberry extract need be protected

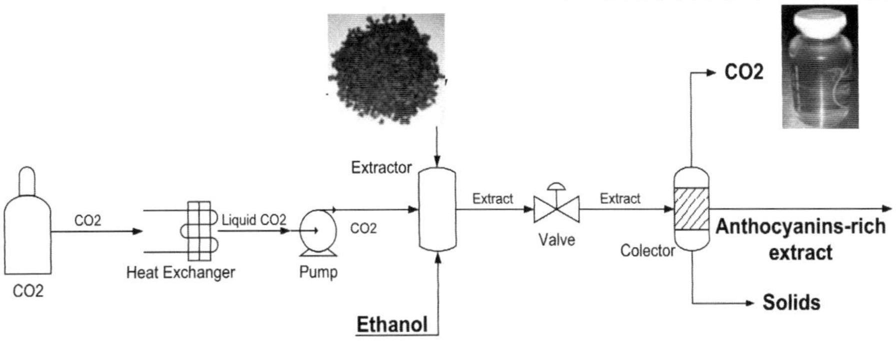

- Presion: 150 – 300 bar
- Temperature: 45 °C

Extraction Process → Sample Preparation → Spray Drying → Chemical Analysis → Modeling and Simulation → Economic Evaluation

3. METHODOLOGY

Delivery System — Designed to encapsulate, protect and release the active ingredients

Blackberry extract

Maltodextrin 20 DE (Solution 6% w/w)

↓

Core in wall solution

Extraction Process → Sample Preparation → Spray Drying → Chemical Analysis → Modeling and Simulation → Economic Evaluation

3. METHODOLOGY

Microencapsulation Method.

Conditions:

Flow rate 0.49 kg/h.

Inlet air temperature:160 and Outlet air temperature: 80ºC

Nozzle atomizer: 0.7mm diameter

Drying flow rate: 0.36 m3/h

Aspirator flow rate: 35 m3/h.

Total Solid content: 30%

Initial phenolic compounds [mg GAE/ 100g]: 267.21

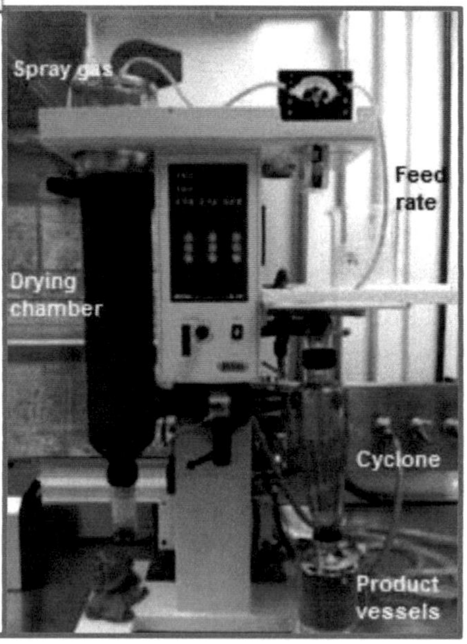

3. METHODOLOGY

Extraction Process → Sample Preparation → Spray Drying → **Chemical Analysis** → Modeling and Simulation → Economic Evaluation

Characterization of the blackberry extract microencapsulated:

Moisture content: (AOAC)

Water activity:(Hygrometer)

Total phenolic compounds (TPC): Folin Ciocalteu method

Total phenolic retention (TPR):

$$TPR = \frac{TPC\ in\ BEM\ (mg/100g)}{TPC\ in\ feed\ (mg/100g)}\ x\ 100$$

Antioxidant capacity: (DPPH method)

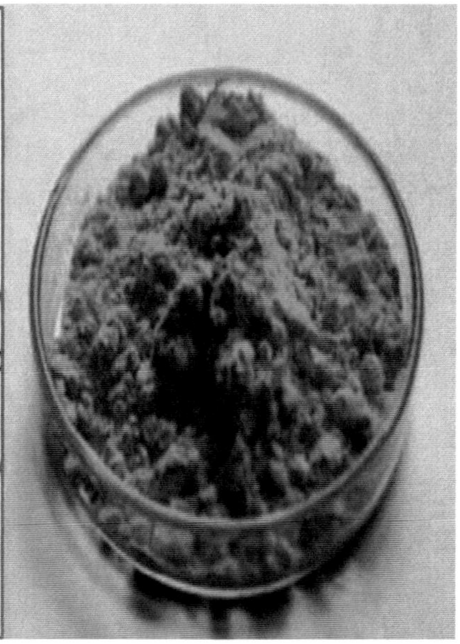

3. METHODOLOGY

Extraction Process → Sample Preparation → Spray Drying → Chemical Analysis → **Modeling and Simulation** → Economic Analysis

Blackberry Extract → Core in Wall solution

Biopolymer → Hydration → Core in Wall solution

Core in Wall solution → Final emulsification → Spray drying → Encapsulated powder

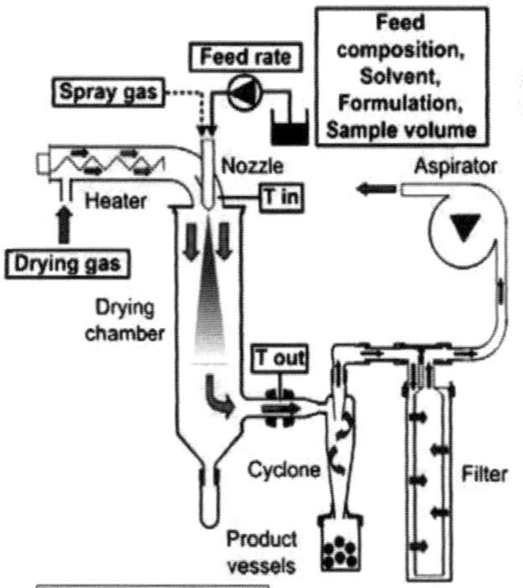

3. METHODOLOGY

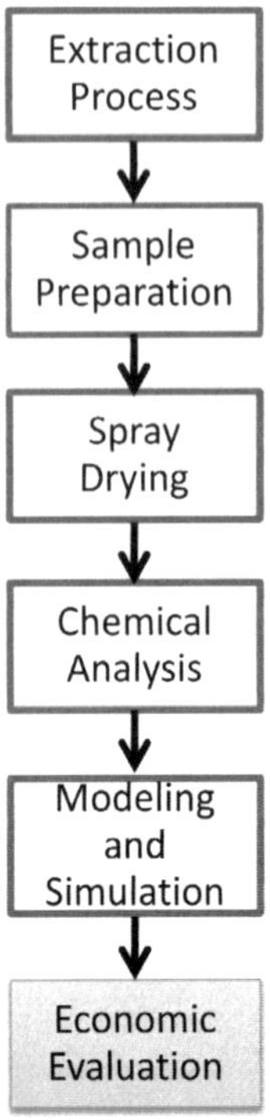

The economic analysis was performed by calculating the production cost per unit of product (e.g. kg). The total production cost considers the total of raw material, utilities, operating labor, maintenance costs, operating charges, plant overhead and general administrative costs. The costs of personnel labor, water and electricity were obtained from typical prices in Colombia.

4. RESULTS AND DISCUSSION

4.1 Chemical Analysis of Blackberry Extract (BE) and Blackberry Extract Microencapsulate [BEM]

Chemical analysis	[BE]	[BEM]
Moisture content [%]	92.47	3.42
Water activity [aw]	0.97	0.26
Phenolic compounds [mg GAE/ 100g]	267.21	253.30
Total phenolic Retention [%]	N.A	94.79
Antioxidant activity DPPH [EC_{50} mg sample /mg DPPH]	21.33	30.23

N.A= Not applicable

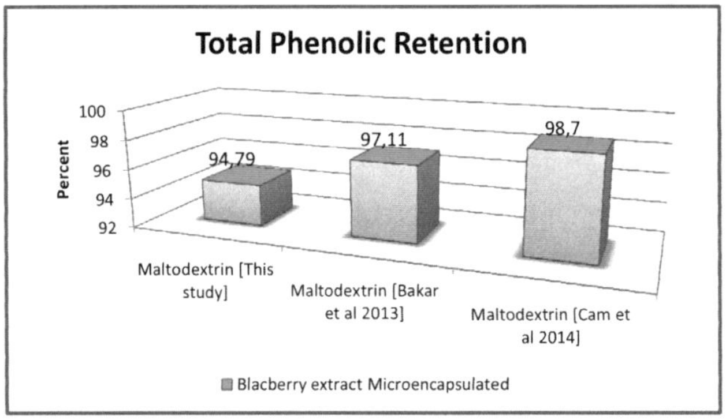

A delivery system should be capable of incorporating an appropriate amount of the active ingredient into the final product

It is important to obtain a high retention of the phenolic compounds within the deliver system until they are released at the required site of action.

4. RESULTS AND DISCUSSION

Chemical Analysis of Blackberry Extract (BE) and Blackberry Extract Microencapsulate [BEM]

Chemical analysis	[BE]	[BEM]
Moisture content [%]	92.47	3.42
Water activity [aw]	0.97	0.26
Phenolic compounds [mg GAE/ 100g]	267.21	253.30
Total phenolic Retention [%]	N.A	94.79
Antioxidant activity DPPH [EC_{50} mg sample /mg DPPH]	21.33	30.23

N.A= Not applicable

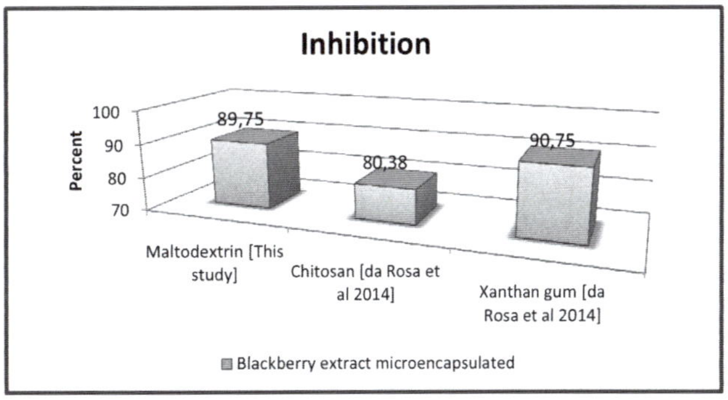

The results from the antioxidant activity studies indicated that the encapsulation procedure had no untoward effect on phenolic compounds.

The phenolic compounds present in the blackberry extract microencapsulated showed a greater antioxidant activity with a high inhibition of 89.75%

4. RESULTS AND DISCUSSION

Modeling and Simulation

Distribution Cost in the production of BEM

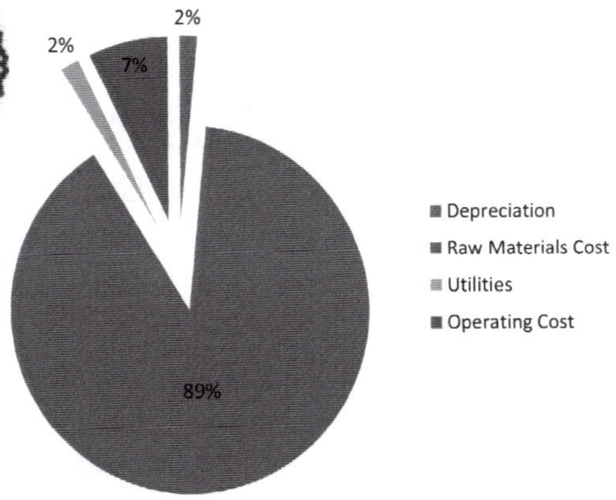

Economic parameters used for BEM estimation cost		
Raw material		
Blackberry Extract[a] (Solvent system CO_2 + EtOH)	USD /kg	7.63
Maltodextrin[c]	USD/ kg	0.38

[a]Previous work
[b]http://www.icis.com/

In this context, it can be observed that the price of the raw material has a great influence on the distribution cost in the production of BEM.
The utilities had a low percentage of the total cost

4. RESULTS AND DISCUSSION

Economic evaluation

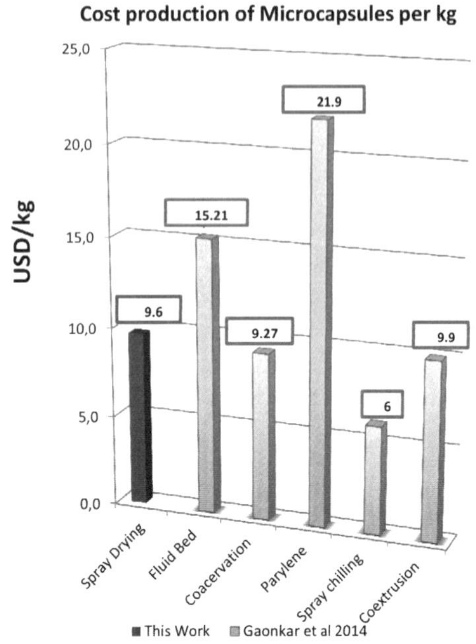

The process economic evaluation largely depends on some factors that must be taken into account before claiming that the BEM process is not economically viable.

Factors:

- Cost and content of the active ingredient to entrap
- Operating conditions
- Water content in the initial feed mixture
- Small scale and high scale

5. CONCLUSIONS

- The BEM obtained by spray drying process had an high retention of 95% of the total phenolic compounds and high inhibition of 89% of free radicals
- The use of polyphenols can not be considered without the protection provided by microencapsulation
- The benefits gained with microencapsulation of blackberry extract should outweigh the additional cost associated with the process of encapsulation

6. ACKNOWLEDGEMENTS

- The authors express their acknowledgments to the Universidad Nacional de Colombia at Manizales
- Research group of Chemical, Catalytic and Biotechnological Processes.

7. REFERENCES

- T. Egan, J.-C. Jacquier, Y. Rosenberg, and M. Rosenberg, "Cold-set whey protein microgels containing immobilised lipid phases to modulate matrix digestion and release of a water-soluble bioactive.," *J. Microencapsul.*, vol. 31, no. 2, pp. 184–92, Jan. 2014.
- T. Egan, J.-C. Jacquier, Y. Rosenberg, and M. Rosenberg, "Cold-set whey protein microgels for the stable immobilization of lipids," *Food Hydrocoll.*, vol. 31, no. 2, pp. 317–324, Jun. 2013.
- C. C. Ferrari, S. P. M. Germer, and J. M. de Aguirre, "Effects of Spray-Drying Conditions on the Physicochemical Properties of Blackberry Powder," *Dry. Technol.*, vol. 30, no. 2, pp. 154–163, Feb. 2012.
- Da Rosa, C. G., Borges, C. D., Zambiazi, R. C., Rutz, J. K., da Luz, S. R., Krumreich, F. D., ... Nunes, M. R. (2014). Encapsulation of the phenolic compounds of the blackberry (Rubus fruticosus). *LWT - Food Science and Technology, 58*(2), 527–533. doi:10.1016/j.lwt.2014.03.042
- Veggi, P. C., Cavalcanti, R. N., & Meireles, M. A. a. (2014). Production of phenolic-rich extracts from Brazilian plants using supercritical and subcritical fluid extraction: Experimental data and economic evaluation. *Journal of Food Engineering, 131*, 96–109. doi:10.1016/j.jfoodeng.2014.01.027
- Santos, D. T., Veggi, P. C., & Meireles, M. A. a. (2010). Extraction of antioxidant compounds from Jabuticaba (Myrciaria cauliflora) skins: Yield, composition and economical evaluation. *Journal of Food Engineering, 101*(1), 23–31. doi:10.1016/j.jfoodeng.2010.06.005
- Anilkumar Gaonkar, Niraj Vasisht, Atul Khare and RS. Microencapsulation in the food industry. A practical Implementation Guide. (Elsevier AP in an imprint of, ed.). San Diego, CA; 2014.

THANK YOU FOR YOUR KIND ATTENTION

Corresponding Author: Carlos Ariel Cardona Alzate.
Km 9 Vía al Aeropuerto La Nubia, Manizales, Caldas
tel. +57 68879400 ext 55354
e-mail: ccardonaal@unal.edu.co
Universidad Nacional de Colombia at Manizales

383760 Liquid-Phase Electrical Discharge Plasmas Rapidly Inactivate Pathogenic and Spoilage Microorganisms in Water

Abstract:

This work seeks to reduce food wastage caused by microbial spoilage through the development of an energy efficient and effective food sterilization technology applicable to the liquid foods manufacturing sector.

We have developed an energy efficient plasma technology that can rapidly achieve 5-log or greater reduction of food spoilage organisms in water. The technology, which creates plasma directly in a liquid, sterilizes microorganisms by combinations of chemical (hydrogen peroxide and oxidative oxygenated species), physical (shockwaves and UV light), and electrical effects.

Through the manipulation of plasma chemistry, we were able to reduce the treatment time for selected microorganisms from 30 min to ~ 50 sec while achieving greater than 5-log reduction, and have noted differences in optimal plasma chemistries for different organisms of interest. The extent of sterilization appears to be determined by the plasma reactor design, input power, plasma electrode material, and electrode polarity. The work also discusses the mechanism of sterilization and investigates the possibility of using the technology to treat fruit juices.

Modeling the Drying Kinetics of Carrot Particles in a Closed Chamber Following Forced Convection and Electric Field Treatment

A.B. Jemai[a], L. Khezami[b], M. HadjKali[a], and E. Vorobiev[c]

a. King Saud University, College of Engineering, Chemical Engineering Dept., P.O. Box 800 Riyadh 11421- KSA.
b. Al Imam Mohammed Ibn Saud University, Faculty of Sciences-Department of Chemistry. P.O.Box: 90950 Riyadh: 11623 Kingdom of Saudi Arabia.
c. University of Technology of Compiegne, Chemical & Industrial Process Engineering Department

The impact of forced convection and Pulsed Electric Field pretreatment on the drying kinetics of grating type carrot particles, placed in a closed chamber, have been studied based on continuous relative humidity measurements ($\Delta\Phi$) of circulating air. The particles were placed in a laboratory-scale closed chamber (30g particles capacity) which can be used for the subsequent pressing operation and drying air circulation. The kinetics of moisture content loss was indirectly deduced from a normalized form of the relative humidity difference between the entering and leaving air streams. Using two thermo-hygrometers, the continuous measurements of relative humidity difference between entering and exiting air were used in a normalized form to study the kinetics of drying for various air speed. The effect of an electric field pre-treatment of the particles was also illustrated using this experimental setup. About 20% reduction in drying time could be achieved following PEF pretreatment.

Keywords: Agro-food particles processing, Relative humidity, Modeling Drying kinetics, Forced convection, Electric fields.

383942 Antimicrobial Peptide Segments from Soy Protein for Use in Food Safety

Abstract:

Antimicrobial peptides (AMPs) kill microbial cells through insertion and damage/permeabilization of the cytoplasmic cell membranes and has applications in food safety. Soy protein may be such an attractive, cost-saving candidate for commercial consideration because the protein subunits have amino acid sequences that contain several α-helix or 3-10 helix domains which possess characteristics of AMPs. A methodology for identification of AMPs from soy protein is proposed. They were identified from soy β-conglycinin (7S) and glycinin (11S) based on (i) number of amino acids, (ii) positive charge, (iii) hydrophobicity and (iv) hydrophobic moment. Explicit solvent molecular Dynamics (MD) simulation was employed to assess the secondary conformation of these peptides in POPC/POPG bilayers to mimic their permeation action on the cell membrane of microorganism. The effects of number of peptides, their orientation and hydrophobic moment on the deformation of membrane and formation of water channel were investigated. A mathematical model for the prediction of minimum inhibitory concentration for deactivation (antimicrobial activity) was proposed accounting for free energy of formation of a pore by an aggregate of peptides. The antimicrobial activity was tested by using synthetic selected peptides against *Listeria monocytogenes* and *E. Coli* and compared with predictions.

James E. Van Wyk P.E.

Process Development for Sustainable Food and Biochemical Production

AIChE Annual Meeting Atlanta GA

19 June 2015

Making Food Safe in Light of New Demands

1. Introduction

Food products are non-Newtonian. This means that there are few if any imperial formulas that can be used to predict heat transfer properties. The most reliable data is from laboratory data and field results. This is compounded by the fact that if a new ingredient is added to an existing product, the properties can change dramatically.

The concern in the design of a food heat exchanger system is to insure that all potentially harmful bacteria are destroyed. When a new ingredient is added it can change the heat properties to an extent that a previous safe product becomes unsafe.

There are numerous examples that can be shown. In this discussion we will focus on two products which have become popular in the current market

1. Orange juice which has now other juices added such as tropical fruits
2. Tomato soup which has products added such a bits of potatoes, carrots and other vegetables

While admittedly there could be many more these two were chosen because of their time in the market as a stand-alone product and how changes drastically changed the way they are processed.

2. Product Particulars

   a. Orange Juice- Orange juice has been on the market for a number of years. The trend since 1990 has been for NFC or not from concentrate. This actually represents the same product as you would get if you squeezed it yourself. Since the orange juice is blended form various types, the consumer gets a consistent product that always tastes the same

      In the early 2000's the demand for single strength orange juice had peaked. In fact the per person consumption of orange juice has dropped. Added is the fact that major producers were attempting to move orange juice from an only breakfast drink to a drink consumed any time of the day.

      To achieve these objectives marketing people added tropical juices to orange juice to add variety of flavor. Many different juices were tried. The one that appeared to be the best taste wise was cranberry.

      From a thermal point cranberry changed the characteristics tremendously. The juice no longer behaved as orange juice with an additive. It behaved as an entirely different product.

      The way we learned was due to spoilage. Orange which is pasteurized has a shelf life of 60-80 days. This product had a shelf life of 10 days. It should be noted that no human became ill from the product. It simply spoiled very quickly

   b. Tomato Soup

      Tomato soup has been available in cans for a long time. The can are retorted (Cooked in can). This actually causes overcooking and can give a burnt flavor to the product. Since 1980 there has been a market demand for a fresher flavor. In order to do this the soup has to be pasteurized and then aseptically filled in a container.

Plain tomato soup sales were either flat or slightly declining. In order to increase volume it was decided to add ingredients. Small diced potatoes and carrots were added.

The issue became how can be we be sure that the potatoes and carrots are adequately heat treated. These products constitute an added mass that not only affects the heat transfer properties but also insurance that they have been heat penetrated to insure safety.

3. Laboratory Results
    a. Orange Juice. We first did a side by side calculation to see what properties might have changed.
        i. Viscosity- No change
        ii. Thermal Conductivity No change
        iii. Heat Capacity-Significant change A decrease of almost 20 %
    b. Tomato Soup We first did a side by side calculation to see what properties might have changed.
        i. Viscosity- A decrease in dynamic viscosity of 29%
        ii. Thermal Conductivity No change
        iii. Heat Capacity-Significant change A decrease of almost 20 %
    c. Laboratory Conclusions-
        i. The design of the heat exchanger had to be modified to meet the change in heat capacity. We either had to add surface or lower the processing rate

ii. With the tomato soup we found that the particles were not heated sufficiently. This meant that the soup had to be exposed to the heating for a longer period. The resonance time went from 10 min to almost 18 minutes. This meant that either we add surface or lower the processing rate,

4. Field Results

   a. Orange Juice-The juice processing rate is closely coupled to the rate of filling. Specifically the processing rate is 100 gpm to coincide with a ½ gallon filling rate of 80 gpm. This meant the only option for us was to increase the heat transfer surface. Since we were using a plate heat exchanger this was accomplished with almost no difficulty. We increase the area to what the laboratory test would indicate. This however proved to be not enough. We still had to increase by an additional 5% to achieve the pasteurization temperature.

   b. Tomato Soup- This proved to be a more difficult fix. We could not increase the surface area due to space limitations and cost. This was being processed in a tubular heat exchanger. We therefore had to slow the flow rate. The actual processing rate was about 30% less. From an engineering point of view this was acceptable. The economics at running at a lower rate proved not the case. Once the product was established in the market an new line as built for the soup

5. Conclusions

   a. Assume you know nothing- You cannot simply add an ingredient and assume that it will behave as before. Since the majority if not all food products are non-Newtonian, laboratory testing must be done each time.

b. <u>Look at economics</u>-Can you add surface area with a minimal amount of cost? Adding this cost will the product still be profitable? In some cases this may kill a project.

c. <u>Never put a product out for public consumption until you are 110 % sure</u>

.

386751 Low Temperature Batch Conversion of Cellulosic Biomass

Abstract:

The project aims at developing small, affordable batch reactors that communities in developing countries can use to produce clean cooking gas from biomass. Our approach extracts clean combustible gases such as methane and hydrogen using low cost locally available catalytic ingredients such as sodium chloride and sodium carbonate. Compared to uncatalyzed biomass conversion, the onset of pyrolysis/gasification shifts to significantly lower temperatures in the presence of catalysts. One possible explanation is that the ions in solution migrate in between cellulosic strands/fibers and disrupt the interstrand hydrogen bonding. Upon drying, the salt nucleates and crystallizes in between the fibers, leading to mechanical strain and loss of structural integrity of the fibers. The recovered biochar can be used as a supplemental source of fuel, while the gas evolved can be collected inside the pressure vessel and connected to a gas burner. This allows the use of a wide range of biomass sources without forcing a community to radically change their way of living, thereby increasing the probability of acceptance of this new method.

387725 Effect of C/N Ratio on Microbial Lipid Production with the Oleaginous Yeast Lipomyces Starkeyi

Abstract:

Neutral lipids have been used as raw materials for the production of biodiesel. Since agricultural and plant-derived neutral lipids are often utilized in food preparations, alternative sources of lipids are being investigated. A number of microorganisms possess the ability to produce significant fractions of their dry weight as neutral lipids. The oleaginous yeast, *Lipomyces starkeyi*, is a promising candidate for industrial production of neutral lipids which will ultimately be converted into biofuels. The yeast grows rapidly on various carbon sources and can produce lipids upto 75% of its dry weight. Lipid accumulation most commonly occurs when the nutrient supply is exhausted in the presence of excess carbon at the stationary phase. Under nutrient limiting conditions, yeast cells accumulate lipids as a means of storage when energy sources are abundantly available and when the cellular mechanisms for the microbial synthesis are active. In this research work, experiments are conducted by varying the C:N ratio in the growth medium, and its effect on cell growth and the accumulation of lipids. The carbon source utilized in this work is laboratory grade sweet potato starch as well as starch from raw sweet potato. All the experiments were conducted in the shake flasks as well as in the fermenters with controlled pH, and temperature conditions. Different feeding strategies will also be studied and discussed with this presentation. Cell mass, starch content and lipid content were analyzed periodically. The cell mass was recorded by weighing the sample before and after drying. Lipid content was measured by Nile red florescence. The supernatant from centrifugation or filtrate was used to measure starch concentration by iodine method.

392434 Recovery of Polyphenol-Enriched Sugarcane Molasses Solution By Ultrafiltration

Abstract:

Blackstrap molasses was diluted with deionized water and purified using ultrafiltration hollow fiber membrane. Evaluation of permeate flux and fouling were compared using polysulfone membrane with MWCO of 500 kDa and 100 kDa. Permeate flux was observed every 5-minute interval and steady time was recorded. The transmembrane pressure was varied from 10 to 15 psi, permeate flux changes, which is dependent on the pressure, was observed and recorded. At the end of ultrafiltration, data on final permeate flux were collected for each transmembrane pressure and for each membrane used. Fouling study was performed at total recycling mode by observing the permeate flux decline with time at transmembrane pressures of 15, 20, and 25 psi. The performance of the two hollow fiber membranes used were compared in terms of ultrafiltration permeate flux and fouling behavior. The membrane rejection with respect to polyphenol content of the initial feed and final permeate were calculated to complete the comparison for the two membranes. The results showed that both the membranes have clarified the permeate solution but the rejection for the 100 kDa membrane was the greatest, around 0.7. For both membranes, fouling data were comparable, but final flux in 100 kDa was much lower by 20% as compared to 500 kDa. Removal efficiency of the unwanted molecules such as potassium and calcium crystals were more significant in 100 kDa membrane. Thus, by evaluating used membranes on the enrichment of polyphenols from molasses, it was concluded that 100 kDa membrane has been more appropriate.

388825 β-1,3-Glucan Production By Metabolically Engineered Agrobacterium Sp. from Cellulosic Sugar

Abstract:

β-Glucans are polysaccharides of D-glucose monomers linked by β-1,3-glycosidic bonds and have been used for immune and heart health functional ingredients in food industry due to their functions as "biological response modifiers" including activating the immune system. In addition, they can be utilized as growth promoting agent of domestic animals without use of antibiotcs in feed industry and lots of companies currently use β-glucan as a functional ingredient in cosmetics. Production of β-1,3-glucans from cellulosic sugar, such as cellobiose, is important as it establishes a more sustainable technology for the synthesis of this commercial product, and allows a cheaper renewable substrate, cellulose, to be used. While cellobiose could be potentially used through a hydrolysis mechanism, in which the glucose dimer is hydrolyzed into two glucose molecules, we have decided to pursue a more energetically favorable mechanism, which relies on cellobiose phosphorylase. Cellobiose phosphorylase is an enzyme specific to cellobiose and catalyzes the phosphorolytic cleavage of cellobiose, generating glucose and glucose-1-phospahte which can be directly used for biosynthesis of sugar nucleotides and polysaccharides including β-1,3-glucan. In order to construct a strain producing β-1,3-glucan from cellobiose, the gene encocding cellobiose phosphorylase (CBP), , was introduced to *Agrobacterium* sp. for improved cellobiose utilization in of this bacterium. Five promoters, including three *Agrobacterium* sp. promoters and two commercial promoters, were tested for expression of CBP. Two engineered strains, with promoters P_{HCE} and $P_{ExoC,}$ produced β-glucan 11.6 g/L and 10.3 g/L, respectively, in shake flask cultivation after 5 days. Compared to that of control strain, the two engineered strains produced 2.5~3.0 folds higher β-glucan. The production level in engineered strains are also higher than that from sucrose (7~8 g/L), the best carbon source for β-glucan production. In 5L jar fermenter cultivation, both engineered strains produced 24.5~25.8 g/L of β-glucan from 50 g/L cellobiose, which is as much as twice of the parent strain. Therefore, engineering of *Agrobacterium* sp. for production of β-glucan via cellobiose phosphorolytic reaction is an effective strategy for high-yield production of β-glucan from cellobiose. The availability of metabolically engineered strain capable of producing β-glucan from cellobiose paves the way for its production from cellulose.

Chemical Imaging of Cheese Microstructure using ATR Microimaging Spectroscopy

Ramazan KIZIL* kizilr@itu.edu.tr
Istanbul Technical University, Chemical Engineering Department. Maslak 34469 Istanbul.

The new avenue for vibrational spectroscopy appears to be chemical imaging of complex materials from their own molecular fingerprint at micrometer scale under a spectrometer-coupled optical microscope through consecutive scanning with an array detector of a large number of small size classical semiconducting MCT detectors. The findings through chemoinformatic images provide two-dimensional spatial distribution of species which has ever revealed before at the molecular level.

Determination of food structure at the microscopic level is expected to significantly contribute to the modern food technology whose prime interest lies in development of controlled processes and operations to preserve, transform and create or destroy food structure. ATR imaging system combines a high refractive index (n=4.01) hemispherical Germanium internal reflection element with a Cassegrain-style reflective objective is capable of producing images at a spatial resolution which is four times greater than transreflectance and transmission techniques.

This talk covers developments in the imaging IR spectroscopy technology and its application on microstructure analysis of a heterogeneous food product, cheese. The spatial distribution of proteins and fats in the cheese matrix was determined from their unique structure related absorption on contrast images at spatial resolution of 1.56 micrometer. It was shown that cheddar cheese, feta cheese and cream cheese all has all different patterns of fat distribution in the texture. In addition the size of fat globules differs greatly from type to type. A sample image, which is obtained from a feta cheese sample, is shown below to illustrate how spatial distribution of chemical constituents appears in the IR chemical imaging maps.

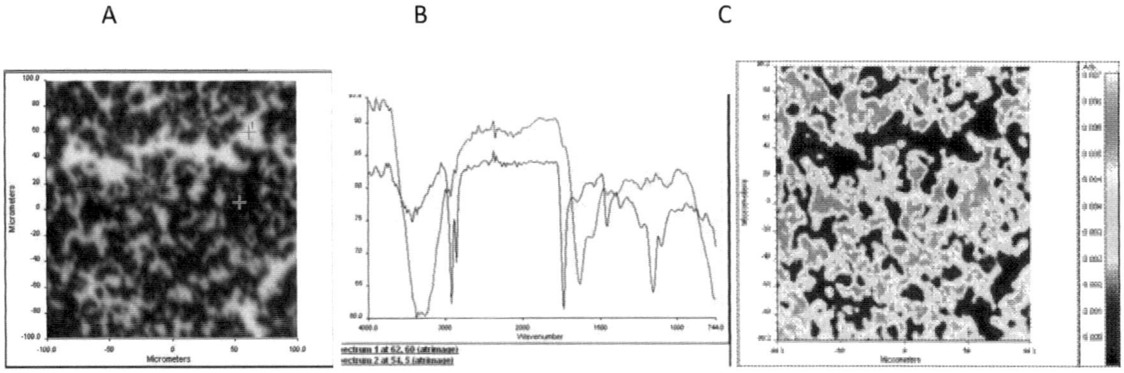

Figure 1. ATR imaging of Feta Cheese A) Raw image based on pattern recognition analysis, B) Each pixel marked on A has full spectral information C) Chemical imaging of cheese based on the amide I vibration absorption to map out spatial distribution of protein in the texture, the red regions depict protein rich sections.

385352 Study on Swine Manure Foaming to Understand Its Major Reason and Its Relations to Industrial Products

Abstract:

In the past few years, a sharp increase of pit-foaming is discovered in swine production facilities. The foams on the swine manure hold significant amount of methane gas, limited the space for manure storage and creates a potential safety hazard. Cases of flash fires and explosions were reported recently in Midwestern states, causing significant assets loss and worker injuries. This research combined with the field sampling and lab simulation to study the foaming mechanism, including manure compositional analysis, microbial community analysis and surfactant identification. Microbial community analysis results show that unlike many foaming cases in wastewater treatment facilities, the manure foaming was not induced by the massive growth of filamentous bacteria. Instead, it is indicated from the lab simulations that the increased concentration of long chain fatty acids together with solid particles in manure solution could initiate the generation of stable foam bubbles. Analysis on solid particles suggests that protein contributes as the major surfactant in manure solution, and further proteomic study will specify which group of protein is mainly involved in this foaming. This research also explored the possibility of some feeding factors that may be related to this foaming phenomenon, like dried distiller grains solubles (DDGS) addition in pig diets and lower digestibility of feed in the pork production. Since DDGS is a major by-product of corn ethanol industry, this will help to improve the industrial process design and provide more applicable products.

Economic analysis of different stabilization methods of antioxidants

Miguel A. Rojas[1], Moshe Rosenberg[2], Carlos A. Cardona[1*]
[1] Biotechnology and Agribusiness Institute, Department of Chemical Engineering
Universidad Nacional de Colombia Sede Manizales, Manizales -Colombia
[2] University of California Davis, Food Science & Technology Department
marojasre@unal.edu.co; mrosenberg@ucdavis.edu; *ccardonaal@unal.edu.co;

[1]

14AIChE
Annual Meeting, Atlanta, GA

1. INTRODUCTION

The growing interest on personal health

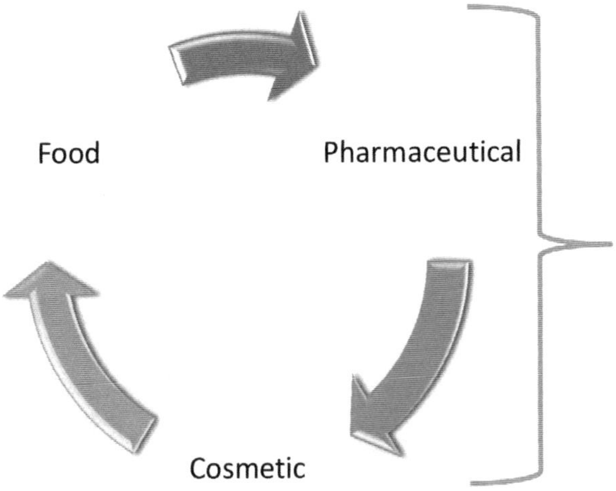

- Flavors
- Antioxidants
- Colors
- Enzymes
- Antimicrobials
- Vitamins
- Minerals
- nutraceuticals

[2]

Research Group in Chemical, Catalytic and Biotechnological Processes
Universidad Nacional de Colombia - Sede Manizales

1. INTRODUCTION

The growing interest on personal health

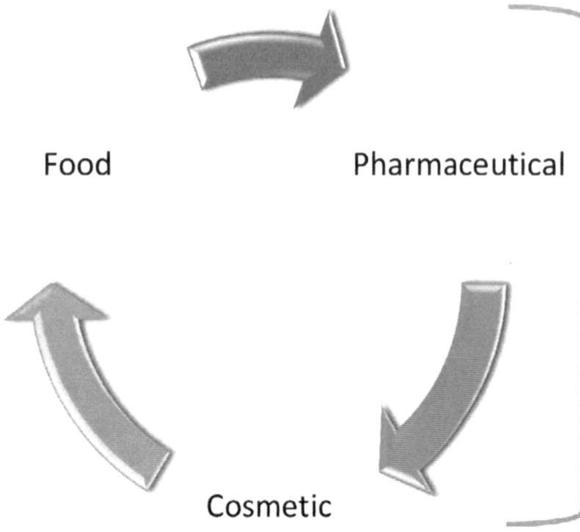

Antioxidants

Unstable

Unplease taste

3

1. INTRODUCTION

After the extraction of active ingredients they need to be immediately protected against biological activity

Food manufacturers

- Antioxidants
- Antimicrobial
- Flavors
- preservatives

Challenges

Overcome by incorporating the active ingredients into some kind of delivery system before it is introduced into the final product

4

1. INTRODUCTION

- A wide variety of microencapsulation techniques are available for utilization in the food industry to encapsulate, protect, and release these active ingredients

Choose the most appropiate system

- Physicochemical properties
- Labelling and legal requirements
- Economic factors

[5]

Research Group in Chemical, Catalytic and Biotechnological Processes
Universidad Nacional de Colombia - Sede Manizales

1. INTRODUCTION

The delivery system should be capable of being economically manufactured from inexpensive ingredients

Economic Factors

- Active ingredients microencapsulated
- Microencapsulation methods
- Encapsulating materials

Prebiotics and probiotics	Herbs and spices
Antioxidants	Carotenoids

[6]

Research Group in Chemical, Catalytic and Biotechnological Processes
Universidad Nacional de Colombia - Sede Manizales

1. INTRODUCTION

The delivery system should be capable of being economically manufactured from inexpensive ingredients

Economic Factors

- Active ingredients microencapsulated
- Microencapsulation methods
- Encapsulating materials

Mechanical	Spray Drying
	Spray Freezing
Chemical	Gelification
	Coacervation
Physical-Chemical	Interfacial polymerization

7

Research Group in Chemical, Catalytic and Biotechnological Processes
Universidad Nacional de Colombia - Sede Manizales

1. INTRODUCTION

The delivery system should be capable of being economically manufactured from inexpensive ingredients

Economic Factors

- Active ingredients microencapsulated
- Microencapsulation methods
- Shell materials

- Proteins
- Simple sugars
- Carbohydrates/gums
- Lipids
- Emulsifiers
- Cellulose

8

Research Group in Chemical, Catalytic and Biotechnological Processes
Universidad Nacional de Colombia - Sede Manizales

2. OBJECTIVE

- The objective of this work perform an economic analysis of the most commonly microencapsulation methods in antioxidants and analyze their economic viability.

9

Research Group in Chemical, Catalytic and Biotechnological Processes
Universidad Nacional de Colombia - Sede Manizales

3. METHODOLOGY

Extract Selection → Stabilization method → Modeling and simulation → Economic analysis

Chemical analysis	Total phenolic Retention [%]	Microencapsulation method
Blackberry Extract Microencapsulated [BEM][a]	94.79	Spray Drying
Bilberry extract microencapsulated [BBA][b]	95	Gelation
Blackberry Extract microencapsuled [BPM][c]	80.52	Complex Coacervation

[a]Previous work
[a]Betz et al. 2012
[C]Zheng et al 2011.

10

Research Group in Chemical, Catalytic and Biotechnological Processes
Universidad Nacional de Colombia - Sede Manizales

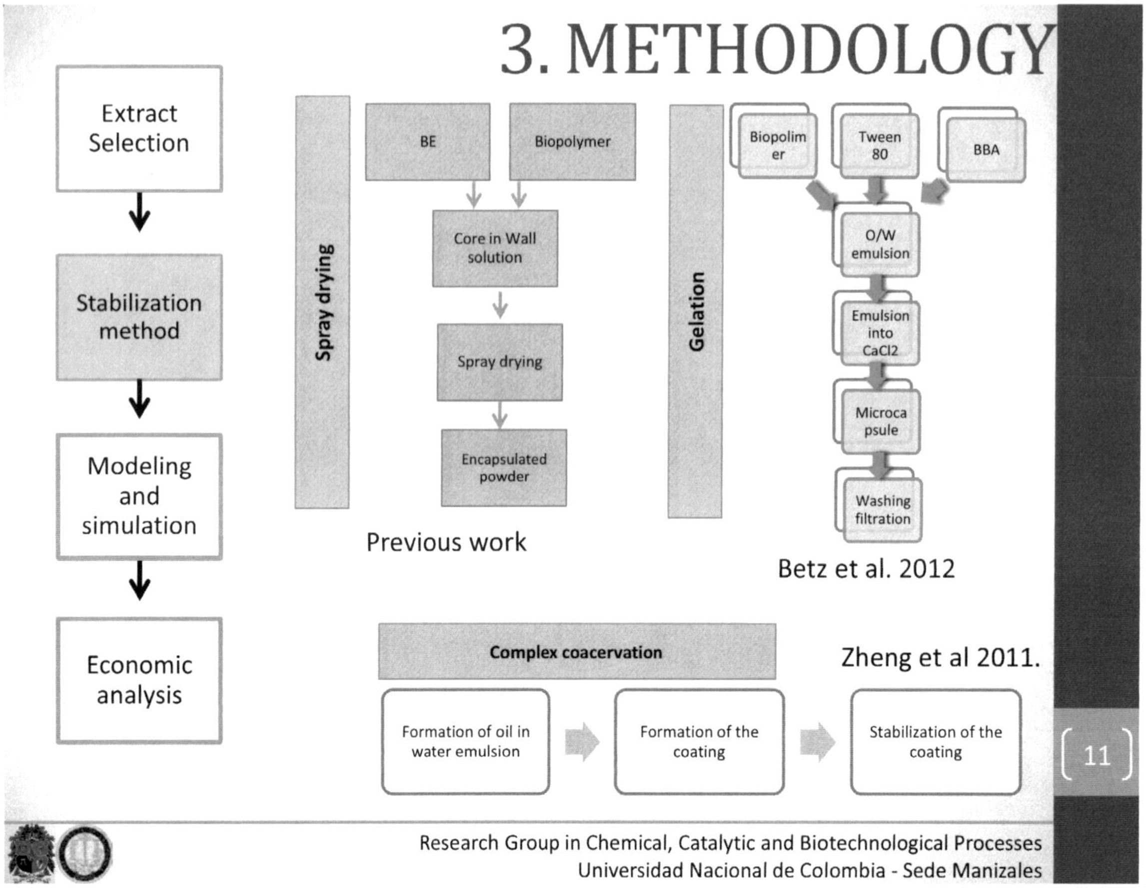
3. METHODOLOGY
Extract Selection
Stabilization method
Modeling and simulation
Economic analysis
Spray drying
BE
Biopolymer
Core in Wall solution
Spray drying
Encapsulated powder
Previous work
Gelation
Biopolim er
Tween 80
BBA
O/W emulsion
Emulsion into CaCl2
Microca psule
Washing filtration
Betz et al. 2012
Complex coacervation
Zheng et al 2011.
Formation of oil in water emulsion
Formation of the coating
Stabilization of the coating
11
Research Group in Chemical, Catalytic and Biotechnological Processes
Universidad Nacional de Colombia - Sede Manizales

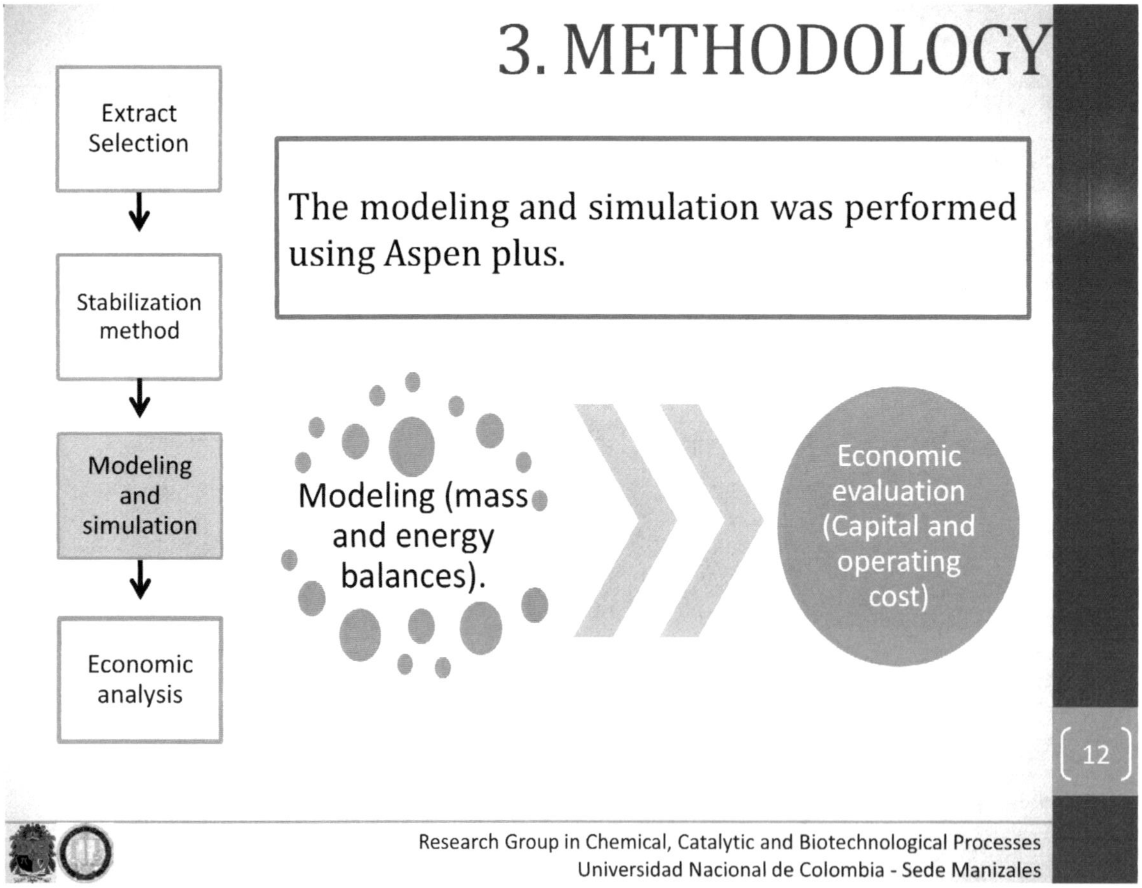
3. METHODOLOGY
Extract Selection
Stabilization method
Modeling and simulation
Economic analysis
The modeling and simulation was performed using Aspen plus.
Modeling (mass and energy balances).
Economic evaluation (Capital and operating cost)
12
Research Group in Chemical, Catalytic and Biotechnological Processes
Universidad Nacional de Colombia - Sede Manizales

3. METHODOLOGY

Extract Selection

↓

Stabilization method

↓

Modeling and simulation

↓

Economic analysis

Economic analysis

The economic analysis was performed by calculating the production cost per unit of product (e.g. kg). The total production cost considers the total of raw material, utilities, operating labor, maintenance costs, operating charges, plant overhead and general administrative costs. The costs of personnel labor, water and electricity were obtained from typical prices in Colombia.

4. RESULTS

Economic parameters used for antioxidant extract microencapsulation [AEM] estimation

Raw material		
Extract[a] (Solvent system CO_2 + EtOH)	USD /kg	7.63
Soybean oil[b]	USD /kg	0.6
Soy lecithin	USD /kg	1.4
Arabic gum[b]	USD/ kg	3.8
Sunflower Oil[b]	USD/ kg	1.3

[a]Previous work
[b]http://www.icis.com/

Modeling and Simulation

In this context, it can be observed that the price of the raw material has a great influence on the distribution cost in the production of microcapsules.
The utilities had a low percentage of the total cost

Distribution Cost in Gelation, Spray drying and coacervation Methods.

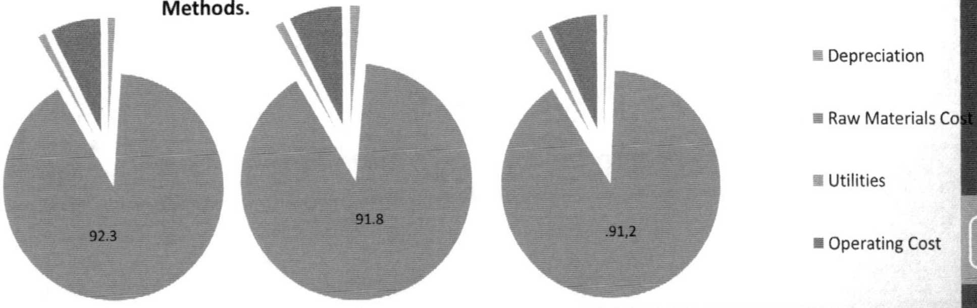

4. RESULTS

Economic evaluation

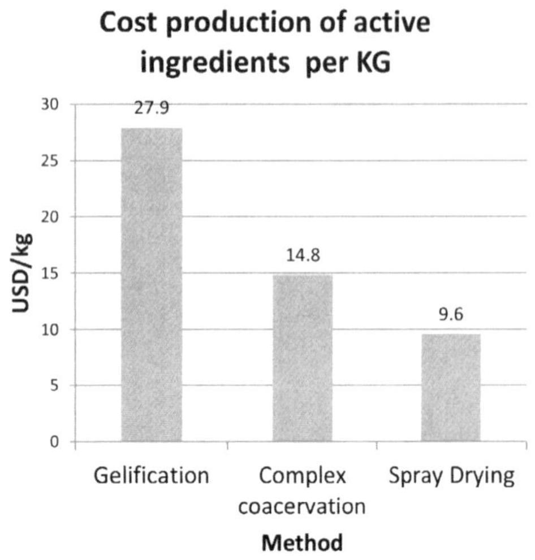

The process economic evaluation largely depends on some factors that must be taken into account before claiming that the microencapsulation method are not economically viable.

Factors:

- Cost and content of the active ingredient to entrap
- Operating conditions
- Water content in the initial feed mixture
- Small scale and high scale

15

Research Group in Chemical, Catalytic and Biotechnological Processes
Universidad Nacional de Colombia - Sede Manizales

5. CONCLUSIONS

- From an economic point of view, the type of microencapsulation method used determine the overall cost of the process, which is an important consideration for the food industry where profit margins are often tight.
- The benefits gained with microencapsulation of active ingredients should outweigh the additional cost associated with the process of encapsulation

16

Research Group in Chemical, Catalytic and Biotechnological Processes
Universidad Nacional de Colombia - Sede Manizales

6. AKNOWLEDGEMDMENTS

- The authors express their acknowledgments to the Universidad Nacional de Colombia sede Manizales
- Research group of Chemical, Catalytic and Biotechnological Processes.

17

Research Group in Chemical, Catalytic and Biotechnological Processes
Universidad Nacional de Colombia - Sede Manizales

7. REFERENCES

- T. Egan, J.-C. Jacquier, Y. Rosenberg, and M. Rosenberg, "Cold-set whey protein microgels containing immobilised lipid phases to modulate matrix digestion and release of a water-soluble bioactive.," *J. Microencapsul.*, vol. 31, no. 2, pp. 184–92, Jan. 2014.
- T. Egan, J.-C. Jacquier, Y. Rosenberg, and M. Rosenberg, "Cold-set whey protein microgels for the stable immobilization of lipids," *Food Hydrocoll.*, vol. 31, no. 2, pp. 317–324, Jun. 2013.
- C. C. Ferrari, S. P. M. Germer, and J. M. de Aguirre, "Effects of Spray-Drying Conditions on the Physicochemical Properties of Blackberry Powder," *Dry. Technol.*, vol. 30, no. 2, pp. 154–163, Feb. 2012.
- Da Rosa, C. G., Borges, C. D., Zambiazi, R. C., Rutz, J. K., da Luz, S. R., Krumreich, F. D., ... Nunes, M. R. (2014). Encapsulation of the phenolic compounds of the blackberry (Rubus fruticosus). *LWT - Food Science and Technology*, *58*(2), 527–533. doi:10.1016/j.lwt.2014.03.042
- Veggi, P. C., Cavalcanti, R. N., & Meireles, M. A. a. (2014). Production of phenolic-rich extracts from Brazilian plants using supercritical and subcritical fluid extraction: Experimental data and economic evaluation. *Journal of Food Engineering*, *131*, 96–109. doi:10.1016/j.jfoodeng.2014.01.027
- Santos, D. T., Veggi, P. C., & Meireles, M. A. a. (2010). Extraction of antioxidant compounds from Jabuticaba (Myrciaria cauliflora) skins: Yield, composition and economical evaluation. *Journal of Food Engineering*, *101*(1), 23–31. doi:10.1016/j.jfoodeng.2010.06.005
- M. Betz, B. Steiner, M. Schantz, J. Oidtmann, K. Mäder, E. Richling, and U. Kulozik, "Antioxidant capacity of bilberry extract microencapsulated in whey protein hydrogels," Food Res. Int., vol. 47, no. 1, pp. 51–57, Jun. 2012.
- L. Zheng, Z. Ding, M. Zhang, and J. Sun, "Microencapsulation of bayberry polyphenols by ethyl cellulose: Preparation and characterization," J. Food Eng., vol. 104, no. 1, pp. 89–95, May 2011.

18

Research Group in Chemical, Catalytic and Biotechnological Processes
Universidad Nacional de Colombia - Sede Manizales

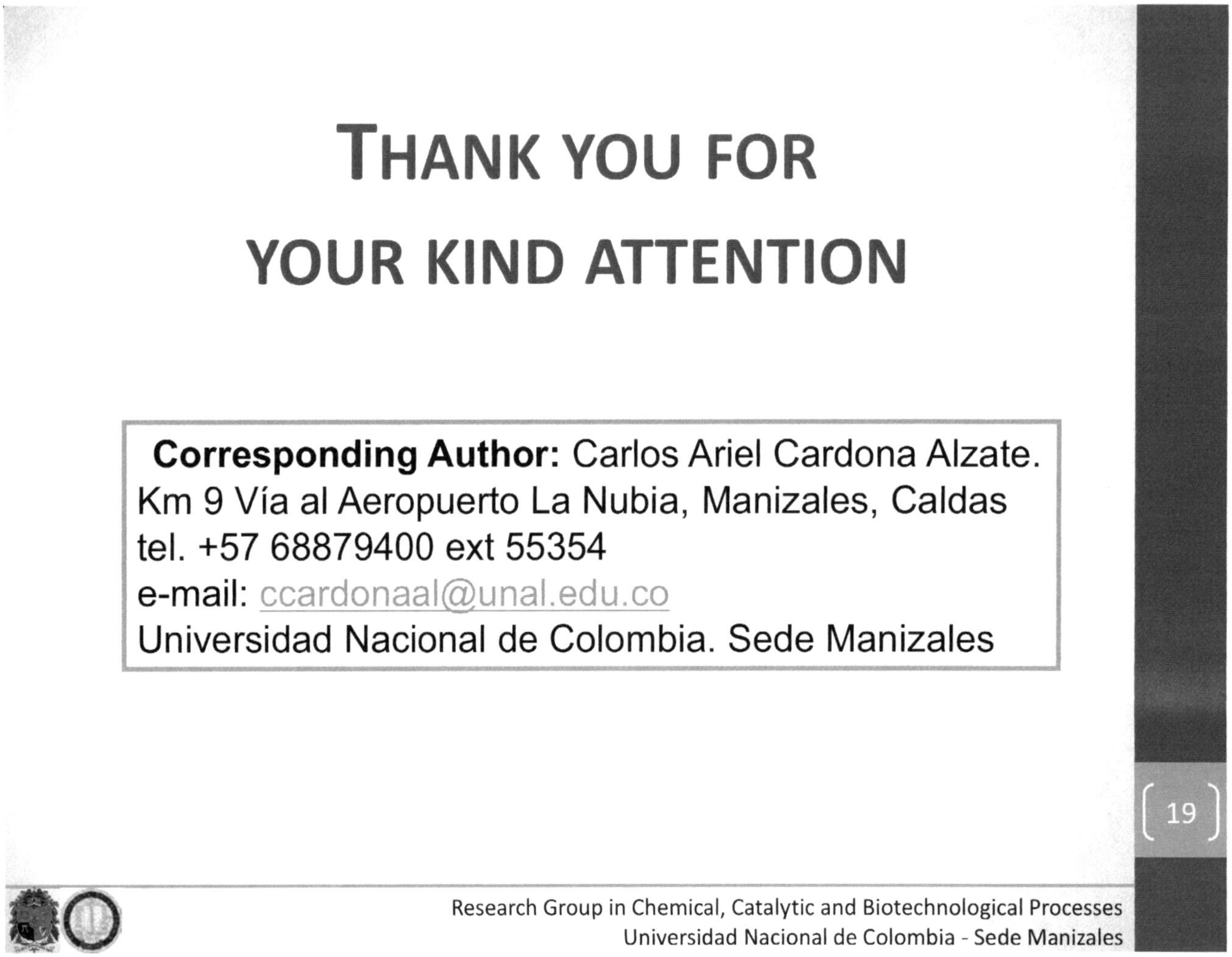
THANK YOU FOR
YOUR KIND ATTENTION
Corresponding Author: Carlos Ariel Cardona Alzate.
Km 9 Vía al Aeropuerto La Nubia, Manizales, Caldas
tel. +57 68879400 ext 55354
e-mail: ccardonaal@unal.edu.co
Universidad Nacional de Colombia. Sede Manizales
19
Research Group in Chemical, Catalytic and Biotechnological Processes
Universidad Nacional de Colombia - Sede Manizales

Preveting Paper Towels from Bacteria Growth Using Selenium Nanoparticles

Qi Wang[1], Thomas J. Webster[2]

[1]Department of Bioengineering Department, [2]Department of Chemical Engineering,
College of Engineering, Northeastern University,
Boston, MA 02115, U.S.A.

Introduction

There are wide spread bacterial contaminations on various paper products, such as paper towels hanging in sink splash zones or those used to clean surfaces, filter papers used in water and air purifying systems, and wrapping used in the food industry, which may lead to the potential spread of bacteria and consequent health concerns. In this study, selenium nanoparticles were coated on normal paper towel surfaces through a quick precipitation method, introducing antibacterial properties to the paper towels. Their effectiveness at preventing biofilm formation was tested in bacterial assays involving *Staphylococcus aureus*, *Pseudomonas aeruginosa*, *Escherichia coli* and *Staphylococcus epidermidis*. The results showed that there were significant and continuous bacteria inhibition with about a 90% reduction from 24 to 72 hours for gram-positive bacteria including *S. aureus* and *S. epidermidis*. The selenium coated paper towels also showed significant inhibition of gram-negative bacteria like *P. aeruginosa* and *E.coli* growth at about 57% and 84%, respectively, after 72 hours of treatment. This study established a promising selenium-based antibacterial strategy to prevent bacterial growth on paper products, which may lead to the avoidance of severe health concerns.

Materials and Methods

Selenium nanoparticles were synthesized and coated on paper towels (MB550A Hand Towel, Tork Advanced, cut into round chips, 7.01mm in diameter) through a simple and quick precipitation reaction, which involves glutathione (reduced form, GSH) (97%, TCI America, Portland, OR) and sodium selenite (99%, Alfa Aesar, Ward Hill, MA) mixed at a 4:1 molar ratio. Selenium nanoparticles were formed and precipitated on the surface of paper towels immediately following the addition of NaOH as visualized by a color change of the reactant solution from colorless to clear red. The paper towel samples were coated for 30s under a 200 rpm (round per minute) agitation to ensure a uniform coating. The coated substrates were rinsed in deionized water five times and soaked in water for 24 hours to remove the free, non-adherent, selenium nanoparticles and remaining reactants. Then, the selenium coated paper towels were left in petri dishes for several days to completely dry before being used in any further experiments or tests. The uncoated samples that were used as a control group in the following experiments were prepared by treating paper towels with the same procedure for selenium coated paper towel samples described above except that sodium hydroxide was not added so that the reaction did not occur.

Four bacterial cell lines were involved in the tests to evaluate the growth of various types of bacteria on selenium coated and uncoated paper towels. Bacteria were propagated in 30 mg/mL tryptic soy broth (TSB) for 20 hours. Each sample was treated separately with 1 mL of the prepared bacterial solutions (10^6 bacteria/ml) and were cultured for either 24, 48 or 72 hours in an incubator (37°C, humidified, 5% CO_2). For those samples that were cultured for 48 and 72 hours, the media was changed with 1 mL of sterile and fresh TSB every 24 hours. After the treatment, the samples were rinsed with a PBS (phosphate buffered saline) solution twice and placed into 1.5ml microfuge tubes with 1ml of PBS. These microfuge tubes were shaken at 3000 rpm for 15 minutes on a vortex mixer to completely release the bacteria attached on the surface into the solution. Solutions with bacteria from each sample were spread on agar plates and bacteria colonies were counted after 20 hours of incubation. All bacterial inhibitory tests were conducted in triplicate and repeated three times. Data were collected and the significant differences were assessed with the probability associated with a one-tailed Student's t-test. Statistical analyses were performed using Microsoft Excel.

Results and Discussion

The results of bacteria assays involving *S. aureus*, *P. aeruginosa*, *E. coli* and *S. epidermidis* showed a high effectiveness for the selenium coated paper towels at inhibiting bacteria growth. As shown in Figure 1, the selenium coatings significantly inhibited the growth of *S. aureus* by about 89% after 24, 48 or 72 hours on the surface of paper towels. Similarly, the selenium coated paper towels had 90.66%, 92.50% and 92.16% less *S. epidermidis* attached (Figure 2) compared with the uncoated paper towels after 24, 48 and 72 hours, respectively. From 24 hours to 48 hours, there was a large increase in the numbers of bacteria, which means the bacteria propagated quickly during this period on paper towels without a selenium coating. However, on the surface of selenium coated paper towels, there was not an obvious increase in the bacteria numbers from 24 hours to 48 hours. In addition, after incubation for 48 hours and 72 hours, the uncoated paper towel was saturated by bacteria, indicating the formation of a matured biofilm, while the selenium coated paper towels still had a significantly reduced number of bacteria after 72 hours. Overall, the numbers of *S. aureus* and *S. epidermidis* on the selenium coated paper towels remained at a low level not increasing from 24 to 48 to 72 hours, indicating successful inhibition to their growth and biofilm formation.

In the bacterial assays with *P. aeruginosa*, the results in Figure 3 also showed a successful inhibition to the growth of *P. aeruginosa* by 55% and 84% after 48 or 72 hours, respectively, on the surface of selenium coated paper towels. The total amount of *P. aeruginosa* increased from 24 hours to 48 hours to 72 hours on paper towels without selenium coatings, while on the surface of selenium coated paper towels, the amount of *P. aeruginosa* decreased from 24 hours to 48 hours and further decreased dramatically from 48 hours to 72 hours. Therefore, the activity of *P. aeruginosa* was successfully inhibited on selenium coated paper towels and the inhibition increased as the time of treatment increased. Moreover, the results of the bacterial inhibitory assays with *E. coli* are shown in Figure 4. The selenium coated paper towels inhibited *E. coli* growth by about 50%~60% compared with the uncoated paper towels after treatment for 24 hours, 48 hours and 72 hours. Overall, compared with the uncoated paper towels, the selenium coated paper towels showed a significant inhibition to the growth of *E. coli* continuously from 24 hours to 72 hours.

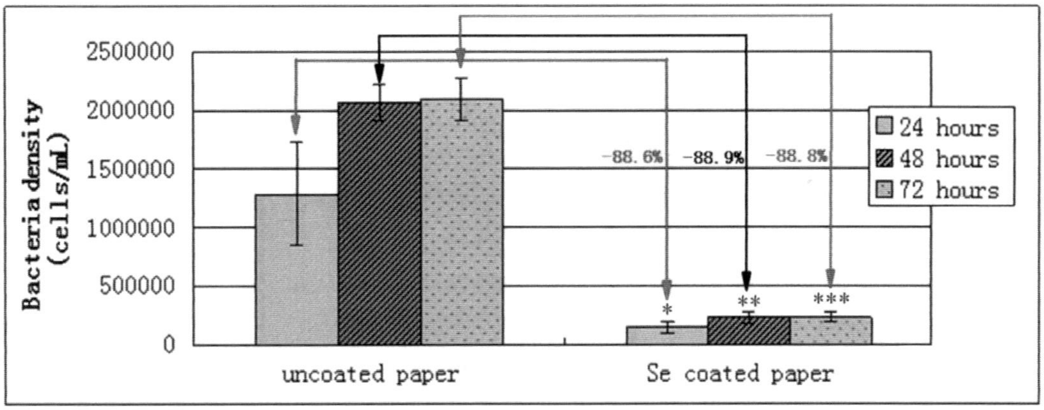

Figure 1. The growth of ***Staphylococcus aureus*** on the surface of paper towels. There was about 90% inhibition on bacteria growth compared with control groups. Data=Mean ± standard deviation, n=3; *, **, ***$p<0.05$ compared with the control group (uncoated paper) after either 24, 48 or 72 hours.

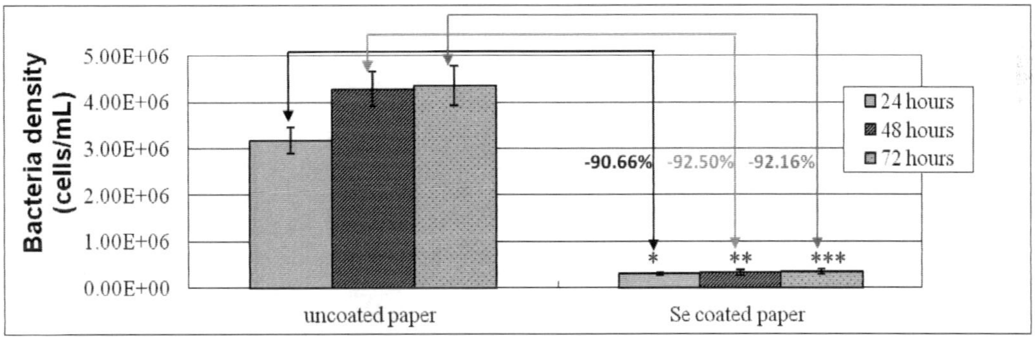

Figure 2. The growth of ***Staphylococcus epidermidis*** on the surface of paper towels. There was more than a 90% inhibition of bacteria growth compared with control groups. Data=mean ± standard deviation, n=3; *, **, ***$p<0.05$ compared with the control group (uncoated paper) after either 24, 48 or 72 hours.

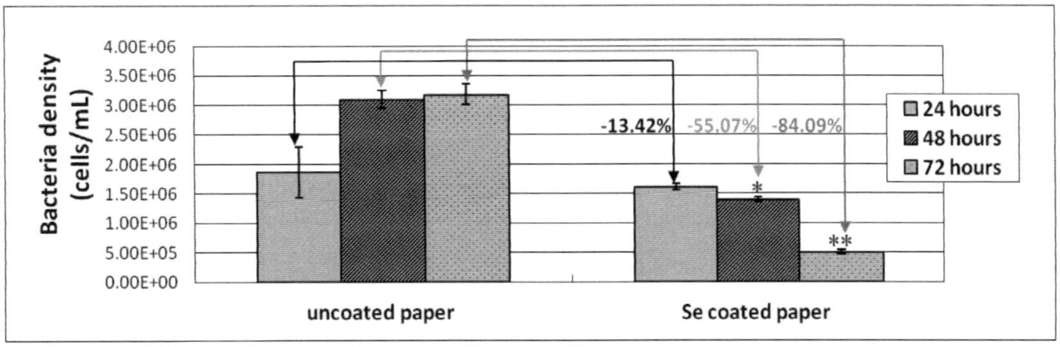

Figure 3.The growth of ***Pseudomonas aeruginosa*** on the surface of paper towels. Data=mean ± standard deviation, n=3; *, **$p<0.05$ compared with the control group (uncoated paper) after either 48 or 72 hours.

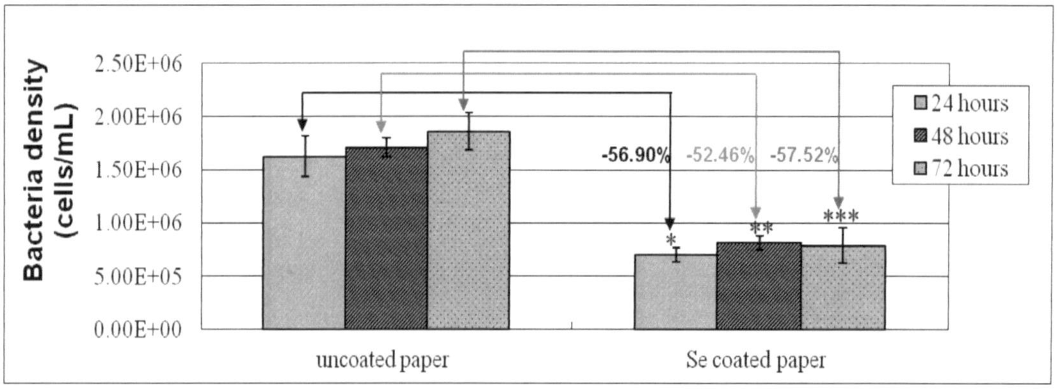

Figure 4. The growth of ***E. coli*** on the surface of paper towels. Data=mean ± standard deviation, n=3; *, **, ***$p<0.05$ compared with the control group (uncoated paper) after either 24, 48 or 72 hours.

Conclusions

Selenium nanoparticles were synthesized and coated on the surface of paper towels through a simple precipitation process which happened within one minute. Impressively, compared with the paper towels without selenium coatings, the selenium coated paper towels significantly inhibited the growth of *S. aureus*, *S. epidermidis*, *P. aeruginosa* and *E.coli* after 24, 48 or 72 hours. There were significant and continuous bacteria inhibition with about a 90% reduction from 24 to 72 hours to the growth of gram-positive bacteria including *S. aureus* and *S. epidermidis*. The selenium coated paper towels showed a relatively lower effectiveness at inhibiting gram-negative bacteria like *P. aeruginosa* and *E.coli* with about a 57% and 84% reduction, respectively, after 72 hours of treatment. Therefore, this study suggested that coating paper products with selenium nanoparticles may be an effective way to decrease various gram-positive and gram-negative bacteria growth on paper products, which might be used for potentially important applications for antimicrobial purposes in the food packaging industry and in clinical environments.

Acknowledgements

The authors thank Northeastern University for funding.

385940 Novel Tools in Genome-Scale Metabolic Flux Modeling to Identify Metabolic Engineering Targets and Predict Microbial Phenotypes

Abstract:

Several new tools have been developed that further enable genome-scale metabolic flux modeling to (i) design metabolic engineering strategies, (ii) select microbial hosts and formulate culture media that favor desired products, and (iii) provide metabolic flux predictions that more closely resemble those determined by ^{13}C-isotopomer metabolic flux analysis (^{13}C-MFA). The Flux Balance Analysis using Flux Ratios (FBrAtio) approach enables the incorporation of specific flux distributions at critical metabolite directly into the stoichiometric matrix of a genome-scale model. Here, we provide detailed instruction on how this is implemented and provide an account on how this method was used to engineer record level cellulose accumulation in the model plant *Arabidopsis thaliana*. Next, the Total Membrane Influx constrained Flux Balance Analysis (ToMI-FBA) tool will be described and used to show that (i) L-valine addition can increase isobutanol production by *Bacillus subtilis*, (ii) cellobiose addition increases ethanol selectivity in *Clostridium acetobutylicum* ATCC 824, and (iii) *B. subtilis* may be an optimal host for artimisinate production. Finally, a novel tool involving minimization of the Total Unconstrained eXchange flux by a Genetic Algorithm (TUX GA) will be described and used to show that biomass equation optimization improves metabolic flux predictions. Here, significantly improved agreement between predicted and ^{13}C-MFA determined fluxes was obtained for autotrophic and heterotrophic growth of *Synechocystis* PCC 6803 using the TUX GA approach. These results were further improved by the incorporation of FBrAtio.

393623 Sustainable Winemaking – Challenges and Opportunities

Abstract:

In the making of wine one can expect to generate 2 gallons to as high as 10 gallons of water for every gallon of wine produced. This water is purely for cleaning/sanitation since only the water contained in the grape can be added to wine. The variation is largely driven by the scarcity/cost of available water, which suggests conservation is intimately tied to economics and the economic conditions are changing at an accelerated rate due to the severity and suddenness of the drought in the western US. A more proactive approach is to consider the problem of water use reduction, treatment (including energy recovery) and synergistic use of the solids streams that exit a typical winery which will be the topic of this talk.

393625 The Challenges of Water Use Reductions and Food Safety in Dairy Processing

Abstract:

The UC Davis Viticulture & Enology Department is developing technologies necessary to realize a self-sustainable, zero carbon winery. Through the use of rainwater capture, filtration and recycling, green cleaning chemistries, Clean-In-Place technology, carbon capture, as well as use of non-carbon energy sources, UCD proposes to develop the world's first energy positive, water positive, and zero carbon winery. Advanced data systems will enable analysis and optimization of processing technologies. Our goal is to develop systems and processes that will enable the food and beverage industry to significantly reduce their water and energy requirements as well as show how process wastewater can be reduced to nearly zero.

THE CHALLENGES OF FOOD SAFETY AND WATER USE REDUCTION IN DAIRY PROCESSING

Montgomery Bohanan and Joe Herrud, Leprino Foods, Denver, CO

Overview

We will discuss the challenge of limiting water usage in conflict with food safety requirements in the dairy industry. As the food industry on a whole strives to deliver the safest food products possible, cleaning requirements and demands continue to increase every year - which is compounded by the fact that the majority of new equipment installations have greater volume demands for cleaning water/fluid. The ability to re-use water is dependent upon and limited by the Federal Pasteurized Milk Ordinance (PMO), and by the infrastructure in place throughout the industrial facilities. With improved equipment designs and the possibility of using enzymes or other cleaning methodologies that are much less water intensive, a significant reduction in water consumption could be made across the dairy manufacturing segment. Chemical engineers working on water reduction around sanitation practices will be vital to simultaneously achieving food safety and water consumption goals within the dairy industry.

Food safety is becoming a greater and greater concern each year as negative health events drive food growers, processors, and manufacturers to work harder to eliminate food-borne risks. The efforts usually result in more cleaning/sanitizing and those steps typically generate more water use. And at the same time, companies are also working hard on reducing the amount of water consumed in their operations to meet sustainability goals or to reduce costs for usage and/or waste water treatment. The availability of potable water in many arid regions (which is more typical in many of the higher dairy production areas of the country) coupled with the ability to treat the additional wastewater generated each day in the sanitation process becomes a balancing act for every organization regarding cleaning and sanitizing and food safety.

Historically, the variables used to achieve the desired level of cleanliness for a clean in place (CIP) system in any unit operation has consisted of:

- Time,
- Temperature,
- Concentration, and
- Mechanical action.

Each variable is dependent upon the others and also upon the soil type and soil load. Let's review the criticality of each of these variables and how they relate to sanitation/food safety.

Time

With enough time, a just about any cleaning strategy may be able to attain the desired soil removal and bacteriological kill. Increased cleaning time could lead to lower temperature requirements or lower cleaning chemical concentration needed. From a food safety standpoint, time is critical in microbial growth and the development of biofilms - so that should also be taken into account. And to maximize capital utilization, the amount of time the equipment is down has to be limited - because time is money. Uptime is vital. Additionally, in most dairy manufacturing processes, sanitation must begin after no longer than every 24 hours of continuous operations (as required in the PMO). Therefore, the ability to sanitize in an efficient amount time will dictate the daily throughput of each manufacturing facility. Typically cleaning time is minimized rather than maximized.

Temperature

Temperature ranges to inhibit bacteriological growth when storing food are typically below 40 degrees F or higher than 140 degrees F. Also, to achieve the necessary kill level as required during the pasteurization step in dairy operations, temperature is time dependent. As an example, raw milk must be held to 161 degrees F for a minimum of 15 seconds for proper pasteurization according to the PMO. So the higher the temperature, the shorter the amount of time needed to achieve desired bacteriological kills. Still, in the thermal process the food flow must pass through the food safety danger zone (between 40 - 140 F), typically twice. These thermal processes for storing, pasteurizing, and otherwise handling food products effect the soil load on equipment surfaces and therefore effect the cleaning and sanitizing process. Soil residues in the food safety danger zone are especially worrisome as they can become microbiological harborages. Typically, cleaning a process requires a temperature minimally 10 F higher than the processing temperature for hot processes. Additionally, the reactivity of cleaning solutions is dependent upon the energy supplied by the temperature of the solution. Temperature can also be dictated by the nature of the soil to be removed; for example in the dairy industry butterfat typically melts at 128 F so wash solutions would exceed that temperature to ease removal of the soil, and usually exceed 140 F. And on the other hand excessively high temperatures can denature or burn on soils, thus making them even harder to remove.

As stated above in the section discussing time needed for the CIP process, there is a balance between the amount of energy utilized to attain desired temperatures and the amount of time the system can be down for sanitation. Typically, cleaning temperatures are limited by energy requirements and equipment tolerance.

Concentration

Water may be the "universal" solvent, but not all soils are water soluble. To improve the efficacy of cleaning and shorten the time required to remove soils, chemicals are used in cleaning solutions dependent upon the soil to be removed and the equipment or surface to be cleaned. Alkaline and/or chlorine solutions have typically been used to remove organic soils, while acid solutions have been used to remove mineral soils by helping to dissolve or emulsify the soil into the cleaning solution.

In the dairy industry, the organic soil load typically consists of butterfat, proteins, and carbohydrates; while the inorganic soils are minerals found in milk such as calcium or minerals deposited from the water source or alkaline cleaners. The amount of cleaning time needed can be shortened with higher concentrations of caustics or acids, but there is a point of diminishing return - when the costs of the higher concentrations outweigh the time saved, and when the higher chemical concentrations begin to damage equipment or prove too costly in the waste water treatment plant.

Sanitizers used in providing bacteriological kill after cleaning are EPA registered pesticides and must be used following label instructions. The efficacy of sanitizing solutions is dependent upon time, temperature and concentration. In a CIP application, the sanitizing solution needs to make contact with the surfaces with enough exposure time to achieve bacteriological kills, and enough volume to ensure complete coverage. For no-rinse sanitizing solutions, a specific concentration range must be maintained to ensure enough chemical to effectively kill microorganisms but still not exceed the maximum level allowed for safe human consumption/exposure. In some instances, this can require significant amounts of potable water to ensure coverage and exposure time on large pieces of equipment or long or complicated line circuits with multiple valve clusters. While critical to ensuring food safety, most sanitizer chemicals prove detrimental to the waste water treatment flora, which can add significant cost to the overall operation.

Typically, chemical concentrations are limited due to usage costs and to waste water treatment and handling costs (COD).

Mechanical Action

The sanitation fluid used in the CIP process must have enough energy to make it to all the surface areas of the equipment, the ability to penetrate biomass or "soil" to be able to chemically dislodge, and be able to scour/remove all residues. The minimum velocity desired in piping systems is typically 5 to 7 feet per second to attain the turbulence needed. Laminar flow will not ensure all soil films are removed from the surfaces. As far as face velocities on other unit operations, the numbers can vary significantly. The flow of the cleaning solution in these applications typically delivers mechanical action through impingement provided by spray devices. Traditionally, supplying improved mechanical action through increased flow rates and thus increased water usage has been the accepted practice. Additionally, as companies have tried to optimize production, the size of equipment and especially piping has increased significantly. The large diameter piping requires large flow rates to fill the entire pipe to ensure coverage of all surfaces plus turbulent flow - when the needed action and contact is only on the inner surface area of the pipe. So, now with cost optimization and water conservation efforts becoming more prevalent, new and improved methods for delivering mechanical action need to be explored.

Additionally, CIP cleaning requires at least two and sometimes several water rinse steps. The initial rinse step in a CIP program is designed to remove as much of the gross soils (and easily removed soils) from the system as possible (targeting 90 - 95% of the soil load). This allows the chemicals in the cleaning solution to work off the hard the remove soils - the films and hardened soils, and can require a large amount of water if a large amount of product is left in the system. Also, there are rinses after each wash step to assist in remove the dissolved soils and to ensure there are no chemical residues left behind. Sometimes these rinses are necessary to ensure incompatible chemical solutions do not inadvertently mix and potentially damage surfaces or equipment. Also, the rinse prior to sanitizing must not only ensure removal of all soil and chemical residues so that microorganisms are not protected from exposure to the sanitizer, but the rinse also is necessary at times to ensure the temperature of the surfaces are appropriate for the sanitizer application.

Even if these variables were always accepted as key to achieving proper sanitation, they are typically incorporated into the design as an afterthought following equipment efficiency, throughput, and capital costs. Cleaning of equipment and processes is often an afterthought to operations, and are therefore difficult to cost or performance optimize. So to achieve sanitation targets in such designed systems and processes, it has been necessary to design equipment to include:

- Higher flow rates and volumes (to increase mechanical action),
- More CIP nozzles and segments (to achieve the necessary coverage),
- Higher temperatures (to maximize solution reactivity), and/or
- Higher concentrations or more effective chemicals (to increase solubility using wetting agents, surfactants, and detergency).

All of which can have a negative impact on water usage, energy usage, and waste water treatment – ultimately the cost of production.

Design for Success

If we were truly optimizing the sanitation process during the design phase of equipment or processes, the following would be considered:

- Minimize residue/biofilm create or left on equipment during operations
    - Keep material deposits to a minimum
        - Eliminate hot spots favorable to biological growth or soil degradation
        - Self cleaning velocities during processing
        - Correct surface finishes or improved surface finished to enhance soil removal
- Make equipment easier to clean
    - Don't have places for bacteria to hide
    - Surface finishes
      No dead spots
    - No ledges
    - Easily and completely drains
    - Accessibility for cleaning
- Ability to remove soils without water (Potential yield improvement)

    - Dry clean
    - Line "pigs" or other air purging devices
    - New methods for purging the system or equipment
- Easy teardown to allow manual removal instead of using water
- Correct style, number, and location of CIP nozzles
    - Design equipment around a spray pattern to minimize wash solution needs instead of adding spray balls to create coverage in an existing piece of equipment
    - Utilize nozzles that maximize coverage and impingement for the volume of water used.
    - Develop new CIP impingement devises.
    - Treat CIP systems and solutions as a limited utility resource and design equipment around those limitations and availability.

Sanitation with Minimal Water or Chemicals

Just pumping more water and additional sanitation chemicals can't always be the answer for proper cleaning. What other options should be investigated to make step changes in the amount of water used?

- Alternative dry cleaning, such as:
    - Dry ice/CO2 residue removal/pretreatment
    - Air pulse or turbulent air blow
- Enzymatic cleaners to break down soils into more readily soluble forms
    - Develop/utilize enzymes that are recoverable and reusable
    - Potential for high concentrations that will not damage equipment while minimizing water usage
- Cleaning chemical recovery and re-use, such as
    - Membrane recovery of caustic/acid
    - Mineral precipitation from acid or of caustic
    - Phase separation from acid or caustic
    - Removal of sanitizing agents from water prior to traditional waste water treatment
    - Associated water recovery from the process listed above
- Different chemical combinations, such as
    - solvent based cleaners
    - acid override washes
    - nanotechology or encapsulation
- Different methods for delivering/applying cleaning chemicals
    - Automated foam application, followed by minimal water impingement
    - Automated and/or robotic brushes and/or spray devices
    - Nanotechnology or encapsulation

The majority of the cleaning techniques/strategies we've discussed are pertaining specifically to processing equipment and CIP systems. There are many other examples of cleaning requirements that each facility must tackle that deal directly with environmental conditions and water consumption at the facility. What are the other practices that can be put into place to address:

- COP (Clean-Out-of-Place) tanks
- Manually cleaning (Foam, Boosted Pressure) and rinsing
- Proper hand cleanliness
- Foot traffic exposure
- Forklift contamination
- Tool contact and cleanliness

Conclusion

Balancing food safety with water reduction is a challenging task within the dairy industry. There are existing best practices we can put in place today to help achieve reduction goals while not sacrificing food safety requirements. But to make a step change for the future, new designs and "out of the box" thinking will be needed. Who will discover the disruptive technology that will allow a reduction in water consumption while maintaining - or improving - plant hygiene and food safety? Who are the creators of "what could be"? You!

393627 Maximizing Water and Energy Efficiency in Food Processing While Minimizing Negative Environmental Impact

Abstract:

The long term health of the food processing industry depends on development of technologies and practices that maximize efficiency of water and energy use while minimizing negative environmental impact. Realizing sustainable food processing hinges on understanding how water and energy use are intertwined throughout food processing facilities down to the level of individual unit operations. The water-energy nexus (WEN) assessment is a framework for auditing all supply and demand points for water and energy within a food processing facility. The WEN analysis provides processors with a complete breakdown of how water and energy use are distributed throughout their processing pipeline and identifies targets for improving efficiency (e.g., by identifying inefficient equipment or opportunities for water and heat recovery). To demonstrate, the results of a WEN analysis from a tomato processing facility are provided. Additionally, opportunities for improving management strategies for food processing solid and liquid wastes are discussed. These include advancing technologies for biofuel production and emerging technologies for agricultural application of food processing organic residues and microbial desalination of wastewater streams.

391639 Land Use Emissions and Environmental Impacts of Agricultural Systems

Abstract:

This poster presents results related to direct land use emissions and fluxes associated with agricultural activities. These emissions are further translated into environmental and climate impacts. The presented results relate to the case study of San Luis Valley in south central Colorado. The emissions (e.g. CO2, N2O etc.) are estimated using DNDC model and impact assessment methods are used to translate them into environmental impacts. We present results related different scenarios such as varying crops, cropping patterns, agricultural practices etc. We also discuss the relevance of these results in the context of regional sustainability.

AUTHOR INDEX